Shi Yu, Léon-Charles Tranchevent, Bart De Moor, and Yves Moreau

# Kernel-based Data Fusion for Machine Learning

# Studies in Computational Intelligence, Volume 345

**Editor-in-Chief**
Prof. Janusz Kacprzyk
Systems Research Institute
Polish Academy of Sciences
ul. Newelska 6
01-447 Warsaw
Poland
*E-mail:* kacprzyk@ibspan.waw.pl

---

Further volumes of this series can be found on our homepage: springer.com

Vol. 321. Dimitri Plemenos and Georgios Miaoulis (Eds.)
*Intelligent Computer Graphics 2010*
ISBN 978-3-642-15689-2

Vol. 322. Bruno Baruque and Emilio Corchado (Eds.)
*Fusion Methods for Unsupervised Learning Ensembles,* 2010
ISBN 978-3-642-16204-6

Vol. 323. Yingxu Wang, Du Zhang, and Witold Kinsner (Eds.)
*Advances in Cognitive Informatics,* 2010
ISBN 978-3-642-16082-0

Vol. 324. Alessandro Soro, Vargiu Eloisa, Giuliano Armano, and Gavino Paddeu (Eds.)
*Information Retrieval and Mining in Distributed Environments,* 2010
ISBN 978-3-642-16088-2

Vol. 325. Quan Bai and Naoki Fukuta (Eds.)
*Advances in Practical Multi-Agent Systems,* 2010
ISBN 978-3-642-16097-4

Vol. 326. Sheryl Brahnam and Lakhmi C. Jain (Eds.)
*Advanced Computational Intelligence Paradigms in Healthcare 5,* 2010
ISBN 978-3-642-16094-3

Vol. 327. Slawomir Wiak and Ewa Napieralska-Juszczak (Eds.)
*Computational Methods for the Innovative Design of Electrical Devices,* 2010
ISBN 978-3-642-16224-4

Vol. 328. Raoul Huys and Viktor K. Jirsa (Eds.)
*Nonlinear Dynamics in Human Behavior,* 2010
ISBN 978-3-642-16261-9

Vol. 329. Santi Caballé, Fatos Xhafa, and Ajith Abraham (Eds.)
*Intelligent Networking, Collaborative Systems and Applications,* 2010
ISBN 978-3-642-16792-8

Vol. 330. Steffen Rendle
*Context-Aware Ranking with Factorization Models,* 2010
ISBN 978-3-642-16897-0

Vol. 331. Athena Vakali and Lakhmi C. Jain (Eds.)
*New Directions in Web Data Management 1,* 2011
ISBN 978-3-642-17550-3

Vol. 332. Jianguo Zhang, Ling Shao, Lei Zhang, and Graeme A. Jones (Eds.)
*Intelligent Video Event Analysis and Understanding,* 2011
ISBN 978-3-642-17553-4

Vol. 333. Fedja Hadzic, Henry Tan, and Tharam S. Dillon
*Mining of Data with Complex Structures,* 2011
ISBN 978-3-642-17556-5

Vol. 334. Álvaro Herrero and Emilio Corchado (Eds.)
*Mobile Hybrid Intrusion Detection,* 2011
ISBN 978-3-642-18298-3

Vol. 335. Radomir S. Stankovic and Radomir S. Stankovic
*From Boolean Logic to Switching Circuits and Automata,* 2011
ISBN 978-3-642-11681-0

Vol. 336. Paolo Remagnino, Dorothy N. Monekosso, and Lakhmi C. Jain (Eds.)
*Innovations in Defence Support Systems – 3,* 2011
ISBN 978-3-642-18277-8

Vol. 337. Sheryl Brahnam and Lakhmi C. Jain (Eds.)
*Advanced Computational Intelligence Paradigms in Healthcare 6,* 2011
ISBN 978-3-642-17823-8

Vol. 338. Lakhmi C. Jain, Eugene V. Aidman, and Canicious Abeynayake (Eds.)
*Innovations in Defence Support Systems – 2,* 2011
ISBN 978-3-642-17763-7

Vol. 339. Halina Kwasnicka, Lakhmi C. Jain (Eds.)
*Innovations in Intelligent Image Analysis,* 2010
ISBN 978-3-642-17933-4

Vol. 340. Heinrich Hussmann, Gerrit Meixner, and Detlef Zuehlke (Eds.)
*Model-Driven Development of Advanced User Interfaces,* 2011
ISBN 978-3-642-14561-2

Vol. 341. Stéphane Doncieux, Nicolas Bredeche, and Jean-Baptiste Mouret(Eds.)
*New Horizons in Evolutionary Robotics,* 2011
ISBN 978-3-642-18271-6

Vol. 342. Federico Montesino Pouzols, Diego R. Lopez, and Angel Barriga Barros
*Mining and Control of Network Traffic by Computational Intelligence,* 2011
ISBN 978-3-642-18083-5

Vol. 343. XXX

Vol. 344. Atilla Elçi, Mamadou Tadiou Koné, and Mehmet A. Orgun (Eds.)
*Semantic Agent Systems,* 2011
ISBN 978-3-642-18307-2

Vol. 345. Shi Yu, Léon-Charles Tranchevent, Bart De Moor, and Yves Moreau
*Kernel-based Data Fusion for Machine Learning,* 2011
ISBN 978-3-642-19405-4

Shi Yu, Léon-Charles Tranchevent, Bart De Moor, and
Yves Moreau

# Kernel-based Data Fusion for Machine Learning

## Methods and Applications in Bioinformatics and Text Mining

Springer

**Dr. Shi Yu**
University of Chicago
Department of Medicine
Institute for Genomics and Systems Biology
Knapp Center for Biomedical Discovery
900 E. 57th St. Room 10148
Chicago, IL 60637
USA
E-mail: shiyu@uchicago.edu

**Dr. Léon-Charles Tranchevent**
Katholieke Universiteit Leuven
Department of Electrical Engineering
Bioinformatics Group, SCD-SISTA
Kasteelpark Arenberg 10
Heverlee-Leuven, B3001
Belgium
E-mail: Leon-Charles.Tranchevent@esat.kuleuven.be

**Prof. Dr. Bart De Moor**
Katholieke Universiteit Leuven
Department of Electrical Engineering
SCD-SISTA
Kasteelpark Arenberg 10
Heverlee-Leuven, B3001
Belgium
E-mail: bart.demoor@esat.kuleuven.be

**Prof. Dr. Yves Moreau**
Katholieke Universiteit Leuven
Department of Electrical Engineering
Bioinformatics Group, SCD-SISTA
Kasteelpark Arenberg 10
Heverlee-Leuven, B3001
Belgium
E-mail: Yves.Moreau@esat.kuleuven.be

ISBN 978-3-642-19405-4 e-ISBN 978-3-642-19406-1

DOI 10.1007/978-3-642-19406-1

Studies in Computational Intelligence ISSN 1860-949X

Library of Congress Control Number: 2011923523



*Typeset & Cover Design:* Scientific Publishing Services Pvt. Ltd., Chennai, India.

Printed on acid-free paper

9 8 7 6 5 4 3 2 1

springer.com

# Preface

The emerging problem of data fusion offers plenty of opportunities, also raises lots of interdisciplinary challenges in computational biology. Currently, developments in high-throughput technologies generate Terabytes of genomic data at awesome rate. How to combine and leverage the mass amount of data sources to obtain significant and complementary high-level knowledge is a state-of-art interest in statistics, machine learning and bioinformatics communities.

To incorporate various learning methods with multiple data sources is a rather recent topic. In the first part of the book, we theoretically investigate a set of learning algorithms in statistics and machine learning. We find that many of these algorithms can be formulated as a unified mathematical model as the Rayleigh quotient and can be extended as dual representations on the basis of Kernel methods. Using the dual representations, the task of learning with multiple data sources is related to the kernel based data fusion, which has been actively studied in the recent five years.

In the second part of the book, we create several novel algorithms for supervised learning and unsupervised learning. We center our discussion on the feasibility and the efficiency of multi-source learning on large scale heterogeneous data sources. These new algorithms are encouraging to solve a wide range of emerging problems in bioinformatics and text mining.

In the third part of the book, we substantiate the values of the proposed algorithms in several real bioinformatics and journal scientometrics applications. These applications are algorithmically categorized as ranking problem and clustering problem. In ranking, we develop a multi-view text mining methodology to combine different text mining models for disease relevant gene prioritization. Moreover, we solidify our data sources and algorithms in a gene prioritization software, which is characterized as a novel kernel-based approach to combine text mining data with heterogeneous genomic data sources using phylogenetic evidence across multiple species. In clustering, we combine multiple text mining models and multiple genomic data sources to identify the disease relevant partitions of genes. We also apply our methods in scientometric field to reveal the topic patterns of scientific publications. Using text mining technique, we create multiple lexical models for more than 8000 journals retrieved from Web of Science database. We also construct multiple interaction graphs by investigating the citations among these journals. These two types of information (lexical /citation) are combined together to automatically construct the structural clustering of journals. According to a systematic benchmark study, in both ranking and clustering problems, the machine learning performance is significantly improved by the thorough combination of heterogeneous data sources and data representations.

The topics presented in this book are meant for the researcher, scientist or engineer who uses Support Vector Machines, or more generally, statistical learning methods. Several topics addressed in the book may also be interesting to computational biologist or bioinformatician who wants to tackle data fusion challenges in real applications. This book can also be used as reference material for graduate courses such as machine learning and data mining. The background required of the reader is a good knowledge of data mining, machine learning and linear algebra.

This book is the product of our years of work in the Bioinformatics group, the Electrical Engineering department of the Katholieke Universiteit Leuven. It has been an exciting journey full of learning and growth, in a relaxing and quite Gothic town. We have been accompanied by many interesting colleagues and friends. This will go down as a memorable experience, as well as one that we treasure. We would like to express our heartfelt gratitude to Johan Suykens for his introduction of kernel methods in the early days. The mathematical expressions and the structure of the book were significantly improved due to his concrete and rigorous suggestions. We were inspired by the interesting work presented by Tijl De Bie on kernel fusion. Since then, we have been attracted to the topic and Tijl had many insightful discussions with us on various topics, the communication has continued even after he moved to Bristol. Next, we would like to convey our gratitude and respect to some of our colleagues. We wish to particularly thank S. Van Vooren, B. Coessen, F. Janssens, C. Alzate, K. Pelckmans, F. Ojeda, S. Leach, T. Falck, A. Daemen, X. H. Liu, T. Adefioye, E. Iacucci for their insightful contributions on various topics and applications. We are grateful to W. Glänzel for his contribution of Web of Science data set in several of our publications.

This research was supported by the Research Council KUL (ProMeta, GOA Ambiorics, GOA MaNet, CoE EF/05/007 SymBioSys, KUL PFV/10/016), FWO (G.0318.05, G.0553.06, G.0302.07, G.0733.09, G.082409), IWT (Silicos, SBO-BioFrame, SBO-MoKa, TBM-IOTA3), FOD (Cancer plans), the Belgian Federal Science Policy Office (IUAP P6/25 BioMaGNet, Bioinformatics and Modeling: from Genomes to Networks), and the EU-RTD (ERNSI: European Research Network on System Identification, FP7-HEALTH CHeartED).

| | |
|---|---|
| Chicago, | Shi Yu |
| Leuven, | Léon-Charles Tranchevent |
| Leuven, | Bart De Moor |
| Leuven, | Yves Moreau |

November 2010

# Contents

# Chapter 1
# Introduction

*When I have presented one point of a subject and the student cannot from it, learn the other three, I do not repeat my lesson, until one is able to.*
*– "The Analects, VII.", Confucius (551 BC - 479 BC) –*

## 1.1 General Background

The history of *learning* has been accompanied by the pace of evolution and the progress of civilization. Some modern ideas of learning (*e.g.*, pattern analysis and machine intelligence) can be traced back thousands of years in the analects of oriental philosophers [16] and Greek mythologies (*e.g., The Antikythera Mechanism* [83]). *Machine learning*, a contemporary topic rooted in computer science and engineering, has always being inspired and enriched by the unremitting efforts of biologists and psychologists in their investigation and understanding of the nature. The Baldwin effect [4], proposed by James Mark Baldwin 110 years ago, concerns the the costs and benefits of *learning* in the context of evolution, which has greatly influenced the development of evolutionary computation. The introduction of *perceptron* and the *backpropagation* algorithm have aroused the curiosity and passion of mathematicians, scientists and engineers to replicate the *biological intelligence* by artificial means. About 15 years ago, Vapnik [81] introduced the support vector method on the basis of kernel functions [1], which has offered plenty of opportunities to solve complicated problems. However, it has also brought lots of interdisciplinary challenges in statistics, optimization theory and applications therein. Though the scientific fields have witnessed many powerful methods proposed for various complicated problems, to compare these methods or problems with the primitive biochemical intelligence exhibited in a unicellular organism, one has to concede that the expedition of human beings to imitate the adaptability and the exquisiteness of *learning*, has just begun.

S. Yu et al.: Kernel-based Data Fusion for Machine Learning, SCI 345, pp. 1–26.
springerlink.com 

## *Learning from Multiple Sources*

Our brains are amazingly adept at *learning from multiple sources*. As shown in Figure 1.1, information travels from multiple senses is integrated and prioritized by complex calculations using biochemical energy at the brain. These types of *integration* and *prioritization* are extraordinarily adapted to environment and stimulus. For example, a student in the auditorium is listening to a talk of a lecturer, the most important information comes from the visual and auditory senses. Though at the very moment the brain is also receiving inputs from the other senses (*e.g.*, the temperature, the smell, the taste), it exquisitely suppresses these less relevant senses and keeps the concentration on the most important information. This *prioritization* also occurs in the senses of the same category. For instance, some sensitive parts of the body (*e.g.*, fingertips, toes, lips) have much stronger representations than other less sensitive areas. For human, some abilities of multiple-source learning are given by birth, whereas some others are established by professional training. Figure 1.2 illustrates a mechanical drawing of a simple component in a telescope, which is composed of projections in several perspectives. Before manufacturing it, an experienced operator of the machine tool investigates all the perspectives in this drawing and combines these multiple 2-D perspectives into a 3-D reconstruction of the component in his/her mind. These kinds of abilities are more advanced and professional than the body senses. In the past two centuries, the communications between the designers and the manufactories in the mechanical industry have been relying on this type of multi-perspective representation and learning. Whatever products either tiny components or giant mega-structures are all designed and manufactured in this

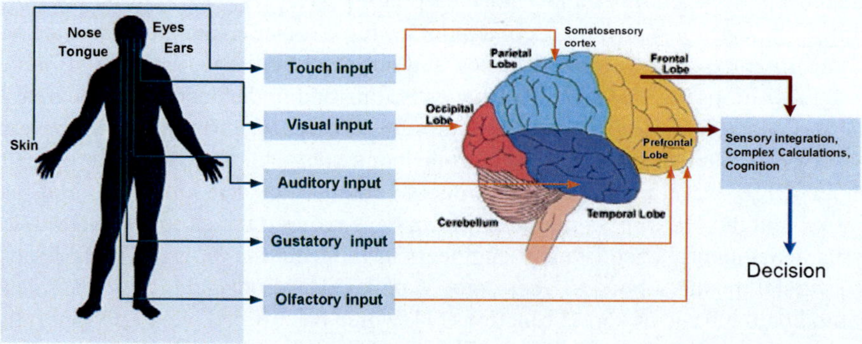


**Fig. 1.1** The decision of human beings relies on the integration of multiple senses. Information travels from the eyes is forwarded to the occipital lobes of the brain. Sound information is analyzed by the auditory cortex in the temporal lobes. Smell and taste are analyzed in the olfactory bulb contained in prefrontal lobes. Touch information passes to the somatosensory cortex laying out along the brain surface. Information comes from different senses is integrated and analyzed at the frontal and prefrontal lobes of the brain, where the most complex calculations and cognitions occur. The figure of human body is adapted courtesy of The Widen Clinic (http://www.widenclinic.com/). Brain figure reproduced courtesy of Barking, Havering & Redbridge University Hospitals NHS Trust (http://www.bhrhospitals.nhs.uk).

manner. Currently, some specialized computer softwares (*e.g.*, AutoCAD, Turbo-CAD) are capable to resemble the human-like representation and reconstruction process using advanced images and graphics techniques, visualization methods, and geometry algorithms. However, even with these automatic softwares, the human experts are still the most reliable sources thus human intervention is still indispensable in any production line.

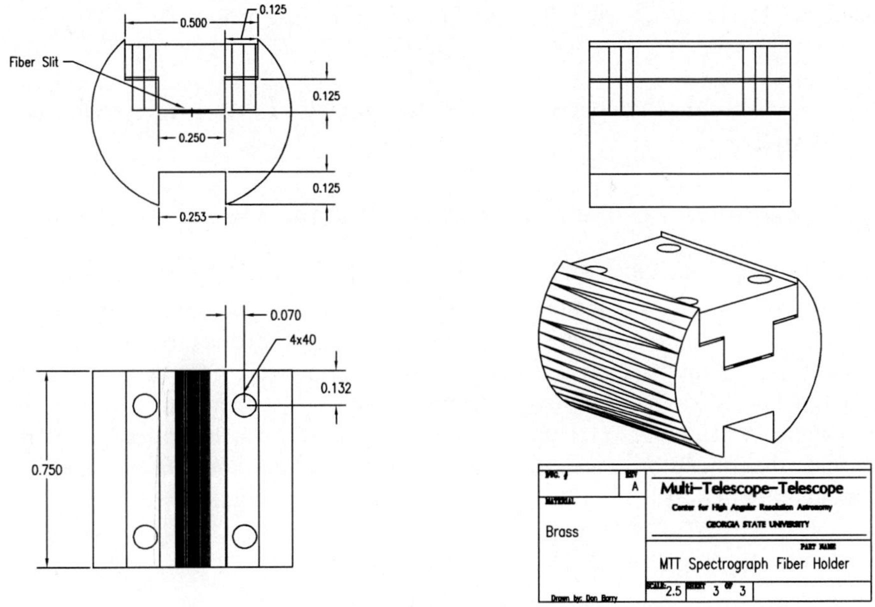


**Fig. 1.2** The method of *multiview orthographic projection* applied in modern mechanical drawing origins from the *applied geometry* method developed by Gaspard Monge in 1780s [77]. To visualize a 3-D structure, the component is projected on three orthogonal planes and different 2-D views are obtained. These views are known as the right side view, the front view, and the top view in the inverse clockwise order. The drawing of the telescope component is reproduced courtesy of Barry [5].

In *machine learning*, we are motivated to imitate the amazing functions of the brain to incorporate *multiple data sources*. Human brains are powerful in learning abstractive knowledge but computers are good at detecting statistical significance and numerical patterns. In the era of information overflow, *data mining* and *machine learning* are indispensable tools to extract useful information and knowledge from the immense amount of data. To achieve this, many efforts have been spent on inventing sophisticated methods and constructing huge scale database. Beside these efforts, an important strategy is to investigate the *dimension* of information and data, which may enable us to coordinate the data ocean into homogeneous threads thus more comprehensive insights could be gained. For example, a lot of data is observed continuously on a same subject at different *time slots* such as the stock market data, the weather monitoring data, the medical records of a patient, and so on. In research of biology, the amount of data is ever increasing due to the advances in *high throughput* biotechnologies. These data sets are often representations of a same group of genomic entities projected in various *facets*. Thus, the idea of incorporating more *facets* of genomic data in analysis may be beneficial, by reducing the noise, as well as improving statistical significance and leveraging the interactions and correlations between the genomic entities to obtain more refined and higher-level information [79], which is known as *data fusion*.

## 1.2 Historical Background of Multi-source Learning and Data Fusion

### *1.2.1 Canonical Correlation and Its Probabilistic Interpretation*

The early approaches of multi-source learning can be dated back to the statistical methods extracting a set of features for each data source by optimizing a dependency criterion, such as *Canonical correlation Analysis* (CCA) [38] and other methods that optimize mutual information between extracted features [6]. CCA is known to be solved analytically as a generalized eigenvalue problem. It can also be interpreted as a probabilistic model [2, 43]. For example, as proposed by Bach and Jordan [2], the maximum likelihood estimates of the parameters $W_1, W_2, \Psi_1, \Psi_2, \mu_1, \mu_2$ of the model illustrated in Figure 1.3:

$$
\begin{aligned}
z &\sim \mathcal{N}(0, I_d), \;\; min\{m_1, m_2\} \geq d \geq 1 \\
x_1 | z &\sim \mathcal{N}(W_1 z + \mu_1, \Psi_1), \;\; W_1 \in \mathbb{R}^{m_1 \times d}, \;\; \Psi_1 \succeq 0 \\
x_2 | z &\sim \mathcal{N}(W_2 z + \mu_2, \Psi_2), \;\; W_2 \in \mathbb{R}^{m_2 \times d}, \;\; \Psi_2 \succeq 0
\end{aligned}
$$

are

$$
\begin{aligned}
\widehat{W}_1 &= \tilde{\Sigma}_{11} U_{1d} M_1 \\
\widehat{W}_2 &= \tilde{\Sigma}_{22} U_{2d} M_2 \\
\widehat{\Psi}_1 &= \tilde{\Sigma}_{11} - \widehat{W}_1 \widehat{W}_1^T \\
\widehat{\Psi}_2 &= \tilde{\Sigma}_{22} - \widehat{W}_2 \widehat{W}_2^T \\
\hat{\mu}_1 &= \tilde{\mu}_1 \\
\hat{\mu}_2 &= \tilde{\mu}_2,
\end{aligned}
$$

where $M_1, M_2 \in \mathbb{R}^{d \times d}$ are arbitrary matrices such that $M_1 M_2^T = P_d$ and the spectral norms of $M_1$ and $M_2$ are smaller than one. The $i$-th columns of $U_{1d}$ and $U_{2d}$ are the first $d$ canonical directions, and $P_d$ is the diagonal matrix of the first $d$ canonical correlations.

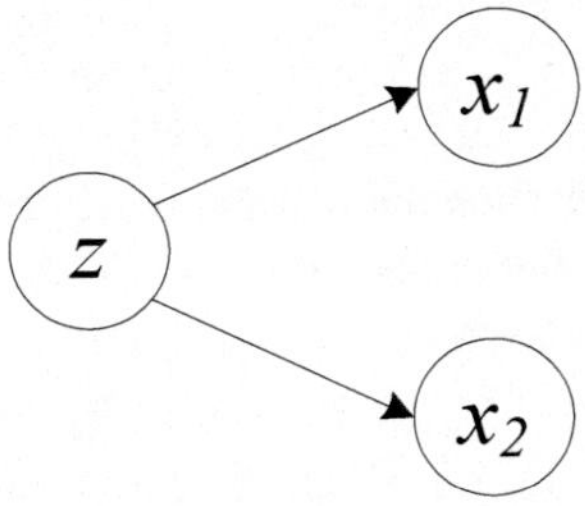


**Fig. 1.3** Graphical model for canonical correlation analysis.

The analytical model and the probabilistic interpretation of CCA enable the use of local CCA models to identify common underlying patterns or same distributions from data consist of independent pairs of related data points. The kernel variants of CCA [35, 46] and multiple CCA are also presented so the common patterns can be identified in the high dimensional space and more than two data sources.

### *1.2.2 Inductive Logic Programming and the Multi-source Learning Search Space*

*Inductive logic programming*(ILP) [53] is a supervised machine learning method which combines automatic learning and first order logic programming [50]. The automatic solving and deduction machinery requires three main sets of information [65]:

1. a set of known vocabulary, rules, axioms or predicates, describing the domain knowledge base $\mathscr{K}$;
2. a set of positive examples $\mathscr{E}^+$ that the system is supposed to describe or characterize with the set of predicates of $\mathscr{K}$;
3. a set of negative examples $\mathscr{E}^-$ that should be excluded from the deducted description or characterization.

Given these data, an ILP solver then finds a set of hypotheses $\mathscr{H}$ expressed with the predicates and terminal vocabulary of $\mathscr{K}$ such that the largest possible subset of $\mathscr{E}^+$ verifies $\mathscr{H}$, and such that the largest possible subset of $\mathscr{E}^-$ does not verify $\mathscr{H}$. The hypotheses in $\mathscr{H}$ are searched in a so-called *hypothesis* space. Different strategies can be used to explore the hypothesis search space (*e.g.*, the *Inductive constraint logic* (ICL) proposed by De Raedt & Van Laer [23]). The search stops when it reaches a clause that covers no negative example but covers some positive examples. At each step, the best clause is refined by adding new literals to its body or applying variable substitutions. The search space can be restricted by a so-called language bias (*e.g.*, a declarative bias used by ICL [22]).

In ILP, data points indexed by the same identifier are represented in various data sources and then merged by an aggregation operation, which can be simply a set union function associated to the inconsistency elimination. However, the aggregation may result in searching a huge space, which in many situations is too computational demanding [32]. Fromont *et al.* thus propose a solution to learn rules independently from each sources; then the learned rules are used to bias a new learning process from the aggregated data [32].

### 1.2.3 Additive Models

The idea of using multiple classifiers has received increasing attentions as it has been realized that such approaches can be more robust (*e.g.*, less sensitive to the tuning of their internal parameters, to inaccuracies and other defects in the data) and be more accurate than a single classifier alone. These approaches are characterized as to learn multiple models independently or dependently and then to learn a unified "powerful" model using the aggregation of learned models, known as the *additive models*. Bagging and boosting are probably the most well known learning techniques based on additive models.

Bootstrap aggregation, or bagging, is a technique proposed by Breiman [11] that can be used with many classification methods and regression methods to reduce the variance associated with prediction, and thereby improve the prediction process. It is a relatively simple idea: many bootstrap samples are drawn from the available data, some prediction method is applied to each bootstrap sample, and then the results are combined, by averaging for regression and simple voting for classification, to obtain the overall prediction, with the variance being reduced due to the averaging [74].

Boosting, like bagging, is a committee-based approach that can be used to improve the accuracy of classification or regression methods. Unlike bagging, which uses a simple averaging of results to obtain an overall prediction, boosting uses a weighted average of results obtained from applying a prediction method to various samples [74]. The motivation for boosting is a procedure that combines the outputs of many "weak" classifiers to produce a powerful "committee". The most popular boosting framework is proposed by Freund and Schapire called "AdaBoost.M1" [29]. The "weak classifier" in boosting can be assigned as any classifier (*e.g.*, when applying the classification tree as the "base learner" the improvements are often dramatic [10]). Though boosting is originally proposed to combine "weak classifiers", some approaches also involve "strong classifiers" in the boosting framework (*e.g.*, the ensemble of Feed-forward neural networks [26][45]).

In boosting, the elementary objective function is extended from a single source to multiple sources through additive expansion. More generally, the basis function expansions take the form

$$f(\mathbf{x}) = \sum_{j=1}^{p} \theta_j b(\mathbf{x}; \gamma_j), \tag{1.1}$$

where $\theta_j$ is the expansion coefficient, $j = 1, ..., p$ is the number of models, and $b(\mathbf{x}; \gamma) \in \mathbb{R}$ are usually simple functions of the multivariate input $\mathbf{x}$, characterized by a set of parameters $\gamma$ [36]. The notion of additive expansions in mono-source can be straightforwardly extended to multi-source learning as

$$f(\mathbf{x}_j) = \sum_{j=1}^{p} \theta_j b(\mathbf{x}_j; \gamma_j), \tag{1.2}$$

where the input $\mathbf{x}_j$ as multiple representations of a data point. The prediction function is therefore given by

$$P(x) = \text{sign}\left(\sum_{j=1}^{p} \theta_j P_j(\mathbf{x}_j)\right), \tag{1.3}$$

where $P_j(\mathbf{x}_j)$ is the prediction function of each single data source. The additive expansions in this form are the essence of many machine learning techniques proposed for enhanced mono-source learning or multi-source learning.

### 1.2.4 Bayesian Networks for Data Fusion

*Bayesian networks* [59] are probabilistic models that graphically encode probabilistic dependencies between random variables [59]. The graphical structure of the model imposes qualitative dependence constraints. A simple example of Bayesian network is shown in Figure 1.4. A directed arc between variables $z$ and $x_1$ denotes conditional dependency of $x_1$ on $z$, as determined by the direction of the arc. The dependencies in Bayesian networks are measured quantitatively. For each variable and its parents this measure is defined using a conditional probability function or a table (*e.g.*, the *Conditional Probability Tables*). In Figure 1.4, the measure of dependency of $x_1$ on $z$ is the probability $p(x_1|z)$. The graphical dependency structure and

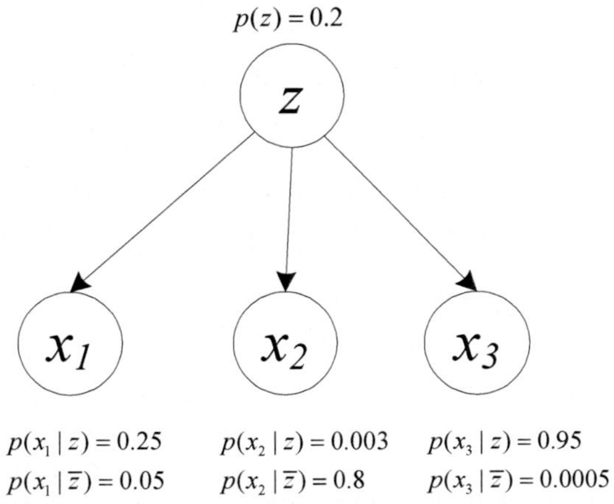


**Fig. 1.4** A simple Bayesian network

the local probability models completely specify a Bayesian network probabilistic model. Hence, Figure 1.4 defines $p(z,x_1,x_2,x_3)$ to be

$$p(z,x_1,x_2,x_3) = p(x_1|z)p(x_2|z)p(x_3|z)p(z). \tag{1.4}$$

To determine a Bayesian network from the data, one need to learn its structure (structural learning) and its conditional probability distributions (parameter learning) [34]. To determine the structure, the sampling methods based on Markov Chain Monte Carlo (MCMC) or the variational methods are often adopted. The two key components of a structure learning algorithm are *searching* for "good" structures and *scoring* these structures. Since the number of model structures is large (super-exponential), a search method is required to decide which structures to score. Even with few nodes, there are too many possible networks to exhaustively score each one. When the number of nodes is large, the task becomes very challenging. Efficient structure learning algorithm design is an active research area. For example, the K2 greedy search algorithm [17] starts with an initial network (possibly with no (or full) connectivity) and iteratively adding, deleting, or reversing an edge, measuring the accuracy of the resulting network at each stage, until a local maxima is found. Alternatively, a method such as simulated annealing guides the search to the global maximum [34, 55]. There are two common approaches used to decide on a "good" structure. The first is to test whether the conditional independence assertions implied by the network structure are satisfied by the data. The second approach is to assess the degree to which the resulting structure explains the data. This is done using a score function which is typically based on approximations of the full posterior distribution of the parameters for the model structure is computed. In real applications, it is often required to learn the structure from incomplete data containing missing values. Several specific algorithms are proposed for structural learning with incomplete data, for instance, the AMS-EM greedy search algorithm proposed by Friedman [30], the combination of evolutionary algorithms and MCMC proposed by Myers [54], the Robust Bayesian Estimation proposed by Ramoni and Sebastiani [62], the Hybrid Independence Test proposed by Dash and Druzdzel [21], and so on.

The second step of Bayesian network building consists of estimating the parameters that maximize the likelihood that the observed data came from the given dependency structure. To consider the uncertainty about parameters $\theta$ in a prior distribution $p(\theta)$, one uses data $d$ to update this distribution, and hereby obtains the posterior distribution $p(\theta|d)$ using Bayes' theorem as

$$p(\theta|d) = \frac{p(d|\theta)p(\theta)}{p(d)}, \quad \theta \in \Theta, \tag{1.5}$$

where $\Theta$ is the parameter space, $d$ is a random sample from the distribution $p(d)$ and $p(d|\theta)$ is likelihood of $\theta$. To maximize the posterior, the Expectation-Maximization (EM) algorithm [25] is often used. The prior distribution describes one's state of knowledge (or lack of it) about the parameter values before examining the data. The prior can also be incorporated in structural learning. Obviously, the choice of the prior is a critical issue in Bayesian network learning, in practice, it rarely happens that the available prior information is precise enough to lead to an exact determination of the prior distribution. If the prior distribution is too narrow it will dominate the posterior and can be used only to express the precise knowledge. Thus, if one has no knowledge at all about the value of a parameter prior to observing the data, the chosen prior probability function should be very broad (non-informative prior) and at relatively to the expected likelihood function.

By far we have very briefly introduced the Bayesian networks. As probabilistic models, Bayesian networks provide a convenient framework for the combination of evidences from multiple sources. The data can be integrated as full integration, partial integration and decision integration [34], which are briefly concluded as follows.

#### Full Integration

In full integration, the multiple data sources are combined at the data level as one data set. In this manner the developed model can contain any type of relationship among the variables in different data sources [34].

#### Partial Integration

In partial integration, the structure learning of Bayesian network is performed separately on each data, which results in multiple dependency structures have only one variable (the outcome) in common. The outcome variable allows joining the separate structures into one structure. In the parameter learning step, the parameter learning proceeds as usual because this step is independent of how the structure was built. Partial integration forbids link among variables of multiple sources, which is similar to imposing additional restrictions in full integration where no links are allowed among variables across data sources [34].

#### Decision Integration

The decision integration method learns a sperate model for each data source and the probabilities predicted for the outcome variable are combined using the weighted coefficients. The weighted coefficients are trained using the model building data set with randomizations [34].

### *1.2.5 Kernel-based Data Fusion*

In the learning phase of Bayesian networks, a set of training data is used either to obtain the point estimate of the parameter vector or to determine a posterior distribution over this vector. The training data is then discarded, and predictions for new inputs are based purely on the learned structure and parameter vector [7]. This approach is also used in nonlinear parametric models such as neural networks [7].

However, there is a set of machine learning techniques keep the training data points during the prediction phase. For example, the Parzen probability model [58], the nearest-neighbor classifier [18], the Support Vector Machines [8, 81], etc. These classifiers typically require a metric to be defined that measures the similarity of any two vectors in input space, as known as the *dual representation*.

### Dural Representation, Kernel Trick and Hilbert Space

Many linear parametric models can be re-casted into an equivalent *dual representation* in which the predictions are also based on linear combinations of a *kernel function* evaluated at the training data points [7]. To achieve this, the data representations are embedded into a vector space $\mathscr{F}$ called the feature space (the Hilbert space) [19, 66, 81, 80]. A key characteristic of this approach is that the embedding in Hilbert space is generally defined implicitly, by specifying an inner product in it. Thus, for a pair of data items, $\mathbf{x}_1$ and $\mathbf{x}_2$, denoting their embeddings as $\phi(\mathbf{x}_1)$ and $\phi(\mathbf{x}_2)$, the inner product of the embedded data $\langle\phi(\mathbf{x}_1), \phi(\mathbf{x}_2)\rangle$ is specified via a kernel function $\mathscr{K}(\mathbf{x}_1, \mathbf{x}_2)$, known as the *kernel trick* or the *kernel substitution* [1], given by

$$\mathscr{K}(\mathbf{x}_1, \mathbf{x}_2) = \phi(\mathbf{x}_1)^T \phi(\mathbf{x}_2). \tag{1.6}$$

From this definition, one of the most significant advantages is to handle symbolic objects (*e.g.*, categorical data, string data), thereby greatly expanding the ranges of problems that can be addressed. Another important advantage is brought by the nonlinear high-dimensional feature mapping $\phi(\mathbf{x})$ from the original space $\mathbb{R}$ to the Hilbert space $\mathscr{F}$. By this mapping, the problems that are not separable by a linear boundary in $\mathbb{R}$ may become separable in $\mathscr{F}$ because according to the VC dimension theory [82], the capacity of a linear classifier is enhanced in the high dimensional space. The dual representation enables us to build interesting extensions of many well-known algorithms by making use of the kernel trick. For example, the nonlinear extension of principal component analysis [67]. Other examples of algorithms extend by kernel trick include kernel nearest-neighbor classifiers [85] and the kernel Fisher Discriminant [51, 52].

### Support Vector Classifiers

The problem of finding linear separating hyperplane on training data consists of $N$ pairs $(\mathbf{x}_1, y_1), ..., (\mathbf{x}_N, y_N)$, with $\mathbf{x}_k \in \mathbb{R}^m$ and $y_k \in \{-1, +1\}$, the optimal separating hyperplane is formulated as

$$\begin{aligned} &\underset{\mathbf{w},b}{\text{minimize}} \quad \frac{1}{2}\mathbf{w}^T\mathbf{w} \\ &\text{subject to} \quad y_k(\mathbf{w}^T\mathbf{x}_k + b) \geq 1, \;\; k = 1, ..., N, \end{aligned} \tag{1.7}$$

where $\mathbf{w}$ is the norm vector of the hyperplane, $b$ is the bias term. The geometry meaning of the hyperplane is shown in Figure 1.5. Hence we are looking for the hyperplane that creates the biggest *margin M* between the training points for class 1 and -1. Note that $M = 2/||\mathbf{w}||$. This is a convex optimization problem (quadratic objective, linear inequality constraints) and the solution can be obtained as via quadratic programming [9].

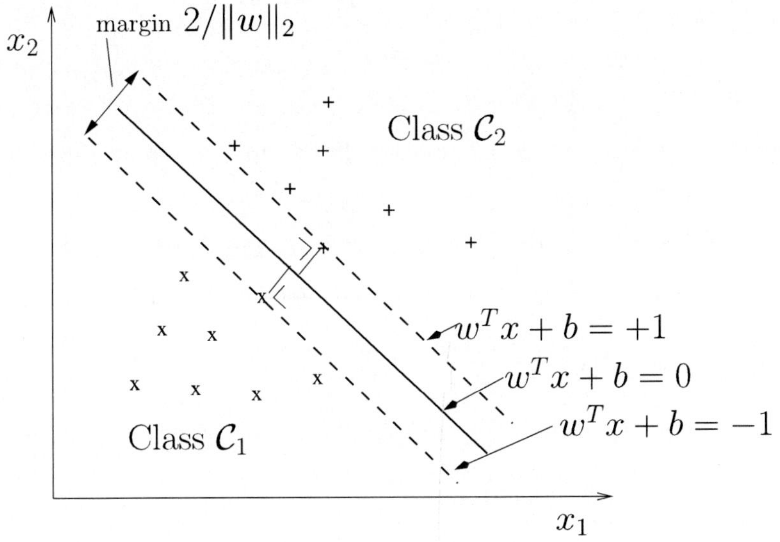


**Fig. 1.5** The geometry interpretation of a support vector classifier. Figure reproduced courtesy of Suykens *et al.* [75].

In most cases, the training data represented by the two classes is not perfectly separable, so the classifier needs to tolerate some errors (allows some points to be on the wrong side of the margin). We define the slack variables $\xi = [\xi_1, ..., \xi_N]^T$ and modify the constraints in (1.7) as

$$\begin{aligned}
&\underset{\mathbf{w},b}{\text{minimize}} \quad \frac{1}{2}\mathbf{w}^T\mathbf{w} && (1.8)\\
&\text{subject to} \quad y_k(\mathbf{w}^T\mathbf{x}_k + b) \geq M(1 - \xi_k), \;\; k = 1, ..., N\\
&\qquad\qquad \xi_k \geq 0, \sum_{k=1}^{N} \xi_k = C, \;\; k = 1, ..., N,
\end{aligned}$$

where $C \geq 0$ is the constant bounding the total misclassifications. The problem in (1.8) is also convex (quadratic objective, linear inequality constraints) and it corresponds to the well known support vector classifier [8, 19, 66, 81, 80] if we replace $\mathbf{x}_i$ with the embeddings $\phi(\mathbf{x}_i)$, given by

$$\underset{\mathbf{w},b,\xi}{\text{minimize}} \quad \frac{1}{2}\mathbf{w}^T\mathbf{w}+C\sum_{k=1}^{N}\xi_k \tag{1.9}$$
$$\text{subject to} \quad y_k[\mathbf{w}^T\phi(\mathbf{x}_k)+b]\geq 1-\xi_k,\ k=1,...,N$$
$$\xi_k\geq 0,\ k=1,...,N.$$

The Lagrange (primal) function is

$$\boxed{\text{P:}}\ \underset{\mathbf{w},b,\xi}{\text{minimize}} \quad \frac{1}{2}\mathbf{w}^T\mathbf{w}+\lambda\sum_{k=1}^{N}-\sum_{k=1}^{N}\alpha_k\left\{y_k\left[\mathbf{w}^T\phi(\mathbf{x}_k^T)+b\right]-(1-\xi_k)\right\}-\sum_{i=1}^{N}\beta_k\xi_k, \tag{1.10}$$

$$\text{subject to} \quad \alpha_k\geq 0,\ \beta_k\geq 0,\ k=1,...,N,$$

where $\alpha_k$, $\beta_k$ are Lagrangian multipliers. The conditions of optimality are given by

$$\begin{cases} \frac{\partial}{\partial \mathbf{w}}=0\rightarrow \mathbf{w}=\sum_{k=1}^{N}\alpha_k y_k\phi(\mathbf{x}_k) \\ \frac{\partial}{\partial \xi}=0\rightarrow 0\leq\alpha_k\leq C,\ k=1,...,N \\ \frac{\partial}{\partial b}=0\rightarrow \sum_{k=1}^{N}\alpha_k y_k=0. \end{cases} \tag{1.11}$$

By substituting (1.11) in (1.10), we obtain the Lagrange dual objective function as

$$\boxed{\text{D:}}\ \underset{\alpha}{\text{maximize}} \quad -\frac{1}{2}\sum_{k,l=1}^{N}\alpha_k\alpha_l y_k y_l\phi(\mathbf{x}_k)^T\phi(\mathbf{x}_l)+\sum_{k=1}^{N}\alpha_k \tag{1.12}$$
$$\text{subject to} \quad 0\leq\alpha_k\leq C,\ k=1,...,N$$
$$\sum_{k=1}^{N}\alpha_k y_k=0.$$

To maximize the dual problem in (1.12) is a simpler convex quadratic programming problem than the primal (1.10). Especially, the Karush-Kuhn-Tucker conditions including the constraints

$$\alpha_k\left\{y_k\left[\mathbf{w}^T\phi(\mathbf{x}_k)+b\right]-(1-\xi_k)\right\}=0,$$
$$\beta_k\xi_k=0,$$
$$y_k\left[\mathbf{w}^T\phi(\mathbf{x}_k)+b\right]-(1-\xi_k)\geq 0,$$

for $k=1,...,N$ characterize the unique solution to the primal and dual problem.

### Support Vector Classifier for Multiple Sources and Kernel Fusion

As discussed before, the additive expansions play a fundamental role in extending mono-source learning algorithms to multi-source learning cases. Analogously, to extend the support vector classifiers on multiple feature mappings, suppose we want to combine $p$ number of SVM models, the output function can be rewritten as

$$f(x) = \sum_{j=1}^{p} \left(\sqrt{\theta_j}\mathbf{w}_j^T \phi_j(\mathbf{x}_k)\right) + b, \tag{1.13}$$

where $\sqrt{\theta_j},\ j = 1,...,p$ are the coefficients assigned to each individual SVM models, $\phi_j(\mathbf{x}_k)$ are multiple embeddings applied to the data sample $\mathbf{x}_k$. We denote

$$\eta = \{\mathbf{w}_j\}_{j=1,...,p}, \text{ and } \psi(\mathbf{x}_k) = \{\sqrt{\theta_j}\phi_j(\mathbf{x}_k)\}_{j=1,...,p}, \tag{1.14}$$

and a pseudo inner product operation of $\eta$ and $\psi(\mathbf{x}_k)$ is thus defined as

$$\eta^T \odot \psi(\mathbf{x}_k) = \sum_{j=1}^{p} \sqrt{\theta_j}\mathbf{w}_j^T \phi_j(\mathbf{x}_k), \tag{1.15}$$

thus (1.13) is equivalently rewritten as

$$f(x) = \eta^T \odot \psi(\mathbf{x}_k) + b, \tag{1.16}$$

Suppose $\theta_j$ satisfy the constraint $\sum_{j=1}^{p} \theta_j = 1$, the new primal problem of SVM is then expressed analogously as

$$\begin{aligned} \underset{\eta,b,\theta,\xi}{\text{minimize}} \quad & \frac{1}{2}\eta^T\eta + C\sum_{k=1}^{N}\xi_k \\ \text{subject to} \quad & y_k\Big[\sum_{j=1}^{p}\sqrt{\theta_j}\mathbf{w}_j^T\phi_j(\mathbf{x}_k) + b\Big] \geq 1 - \xi_k,\ k = 1,...,N \\ & \xi_k \geq 0,\ k = 1,...,N \\ & \theta_j \geq 0,\ \sum_{j=1}^{p}\theta_j = 1,\ j = 1,...,p. \end{aligned} \tag{1.17}$$

Therefore, the primal problem of the additive expansion of multiple SVM models in (1.17) is still a primal problem of SVM. However, as pointed out by Kloft *et al.* [44], the inner product $\sqrt{\theta_j}\mathbf{w}_j$ makes the objective (1.17) non-convex so it needs to be replaced as a variable substitution $\hat{\eta}_j = \sqrt{\theta_j}\mathbf{w}_j$, thus the objective is rewritten as

$$\begin{aligned} \boxed{\text{P:}} \quad \underset{\hat{\eta},b,\theta,\xi}{\text{minimize}} \quad & \frac{1}{2}\sum_{j=1}^{p}\hat{\eta}_j^T\hat{\eta}_j + C\sum_{k=1}^{N}\xi_k \\ \text{subject to} \quad & y_k\left[\sum_{j=1}^{p}\left(\hat{\eta}_j^T\phi_j(\mathbf{x}_k)\right) + b\right] \geq 1 - \xi_k,\ k = 1,...,N \\ & \xi_k \geq 0,\ k = 1,...,N \\ & \theta_j \geq 0,\ \sum_{j=1}^{p}\theta_j = 1,\ j = 1,...,p, \end{aligned} \tag{1.18}$$

where $\hat{\eta}_j$ are the scaled norm vectors $\mathbf{w}$ (multiplied by $\sqrt{\theta_j}$) of the separating hyperplanes for the additive model of multiple feature mappings. In the formulations mentioned above we assume that multiple feature mappings are created on a mono-source problem. It is analogous and straightforward to extend the same objective for multi-source problems. The investigation of this problem has been pioneered by Lanckriet *et al.* [47] and Bach *et al.* [3] and the solution is established in the dual representations as a min-max problem, given by

$$\boxed{\text{D:}}\quad \underset{\theta}{\text{minimize}}\ \underset{\alpha}{\text{maximize}}\ -\frac{1}{2}\sum_{k,l=1}^{N}\alpha_k\alpha_l y_k y_l \sum_{j=1}^{p}\left(\theta_j K_j(\mathbf{x}_k,\mathbf{x}_l)\right)+\sum_{k=1}^{N}\alpha_k \quad (1.19)$$

$$\text{subject to}\quad 0 \geq \alpha_k \geq C,\ k=1,...,N$$

$$\sum_{k=1}^{N}\alpha_k y_k = 0,$$

$$\theta_j \geq 0,\ \sum_{j=1}^{p}\theta_j = 1,\ \ j=1,...,p,$$

where $K_j(\mathbf{x}_k,\mathbf{x}_l)$ represents the kernel matrices, $\mathscr{K}_j(\mathbf{x}_k,\mathbf{x}_l) = \phi_j(\mathbf{x}_k)^T\phi_j(\mathbf{x}_l),\ j = 1,...,p$ are the kernel tricks applied on multiple feature mappings. The symmetric, positive semidefinite kernel matrices $K_j$ resolve the heterogeneities of genomic data sources (*e.g.*, vectors, strings, trees, graphs) such that they can be merged additively as a single kernel. Moreover, the non-uniform coefficients of kernels $\theta_j$ leverage the information of multiple sources adaptively. The technique of combining multiple support vector classifiers in the dual representations is also called *kernel fusion*.

### Loss Functions for Support Vector Classifiers

In Support Vector Classifiers, there are many criteria to assess the quality of the target estimation based on observations during the learning. These criteria are represented as different *loss functions* in the primal problem of Support Vector Classifiers, given by

$$\underset{\mathbf{w}}{\text{minimize}}\ \ \frac{1}{2}\mathbf{w}^T\mathbf{w}+\lambda\sum_{k=1}^{N}L[y_k,f(\mathbf{x}_k)], \quad (1.20)$$

where $L[y_k,f(\mathbf{x}_k)]$ is the *loss function* of class label and prediction value penalizing the objective of the classifier. The examples shown above are all based on a specific loss function called *hinge* loss as $L[y_k,f(\mathbf{x}_k)] = |1-y_k f(\mathbf{x}_k)|_+$, where the subscript "+" indicates the positive part of the numerical value. The loss function is also related to the *risk* or *generalization error*, which is an important measure of the goodness of the classifier. The choice of the loss function is a non-trivial issue relevant to estimating the joint probability distribution $p(\mathbf{x},\mathbf{y})$ on the data $\mathbf{x}$ and its label **y**, which is general unknown because the training data only gives us an incomplete knowledge of $p(\mathbf{x},\mathbf{y})$. Table 1.1 presents several popular loss functions adopted in Support Vector Classifiers.

**Table 1.1** Some popular loss functions for Support Vector Classifiers

| Loss Function | $L[y, f(\mathbf{x})]$ | Classifier name |
|---|---|---|
| Binomial Deviance | $log[1+e^{-yf(\mathbf{x})}]$ | logistic regression |
| Hinge Loss | $\|1-yf(\mathbf{x})\|_+$ | SVM |
| Squared Error | $[1-yf(\mathbf{x})]^2$ (equality constraints) | LS-SVM |
| $L_2$ norm | $[1-yf(\mathbf{x})]^2$ (inequality constraints) | 2-norm SVM |
| Huber's Loss | $\begin{cases} -4yf(\mathbf{x}), & yf(\mathbf{x}) < -1 \\ [1-yf(x)]^2, & \text{otherwise} \end{cases}$ | |

### Kernel-based Data Fusion: A Systems Biology Perspective

The kernel fusion framework has been originally proposed to solve the classification problems in computational biology [48]. As shown in Figure 1.6, this framework provides a global view to reuse and integrate information in biological science at the systems level. Our understanding of biological systems has improved dramatically due to decades of exploration. This process has been accelerated even further during the past ten years, mainly due to the genome projects, new technologies such as microarray, and developments in proteomics. These advances have generated huge amounts of data describing biological systems from different aspects [92]. Many centralized and distributed databases have been developed to capture information about sequences and functions, signaling and metabolic pathways, and protein structure information [33]. To capture, organize and communicate this information, markup languages have also been developed [40, 69, 78]. At the knowledge level, successful biological knowledge integration has been achieved at in ontological commitments thus the specifications of conceptualizations are explicitly defined and reused to the broad audience in the field. Though the bio-ontologies have been proved very useful, currently their inductions and constructions are still relied heavily on human curations and the automatic annotation and evaluation of bio-ontolgoies is still a challenge [31]. On one hand, the past decade has seen the emergent text mining technique filling many gaps between data exploration and knowledge acquisition and helping biologists in their explorative reasonings and predictions. On the other hand, the adventure to propose and evaluate hypothesis automatically in machine science [28] is still ongoing, the expansion of the human knowledge now still relies on the justification of hypothesis in new data with existing knowledge. On the boundary to accept or to reject the hypothesis, biologists often rely on statistical models integrating biological information to capture both the static and dynamic information of a biological system. However, modeling and integrating this information together systematically poses a significant challenge, as the size and the complexity of the data grow exponentially [92]. The topics to be discussed in this book belong to the algorithmic modeling culture (the opposite one is the data modeling culture, named by Leo Breiman [12]). All the effort in this book starts with an algorithmic objective; there is few hypothesis and assumption about the data; the generalization from training data to test data relies on the *i.i.d.* assumption in machine learning. We consider the data being generated by a complex and unknown black box modeled by Support Vector Machines with an input $\mathbf{x}$ and an output $y$. Our goal is then to find a function $f(\mathbf{x})$ —- an algorithm that operates on $\mathbf{x}$ to predict the response $y$. The black box is then validated and adjusted in terms of the predictive accuracy.

Integrating data using Support Vector Machines (kernel fusion) is featured by several obvious advantages. As shown in Figure 1.6, biological data has diverse structures, for example, the high dimensional expression data, the sparse protein-protein-interaction data, the sequence data, the annotation data, the text mining data, and so on. The main advantage is that the data heterogeneity is rescued by the use of kernel trick [1], where data who has diverse data structures is all transformed into kernel matrices with the same size. To integrate them, one could follow the classical additive expansion strategy of machine learning to combine them linearly, moreover, to leverage the effect of information sources with different weights. Apart from the simple linear integration, one could also integrate the kernels geometrically or combine them in some specific subspaces. These nonlinear integration methods of kernels have attracted many interests and have been discussed actively in recent machine learning conferences and workshops. The second advantage of kernel fusion lies in its open and extendable framework. As known, Support Vector Machine is compatible to many classical statistical modeling algorithms therefore these algorithms can all be straightforwardly extended by kernel fusion. In this book we will address some machine learning problems and show several real applications based on kernel fusion, for example, novelty detection, clustering, classification, canonical correlation analysis, and so on. But this framework is never restricted to the examples presented in the book, it is applicable to many other problems as well. The third main advantage of the kernel fusion framework is rooted in convex optimization theory, which is a field full of revolutions and progresses. For example, in the past two decades, the convex optimization problems have witnessed contemporary breakthroughs such as interior point methods [56, 72] and thus have being solved more and more efficiently. The challenge to solve very large scale optimization problems using parallel computing and could computing have intrigued people many years. As an open framework, kernel fusion based statistical modeling can benefit from the new advances in the joint field of mathematics, super-computing and operational researches in a very near future.

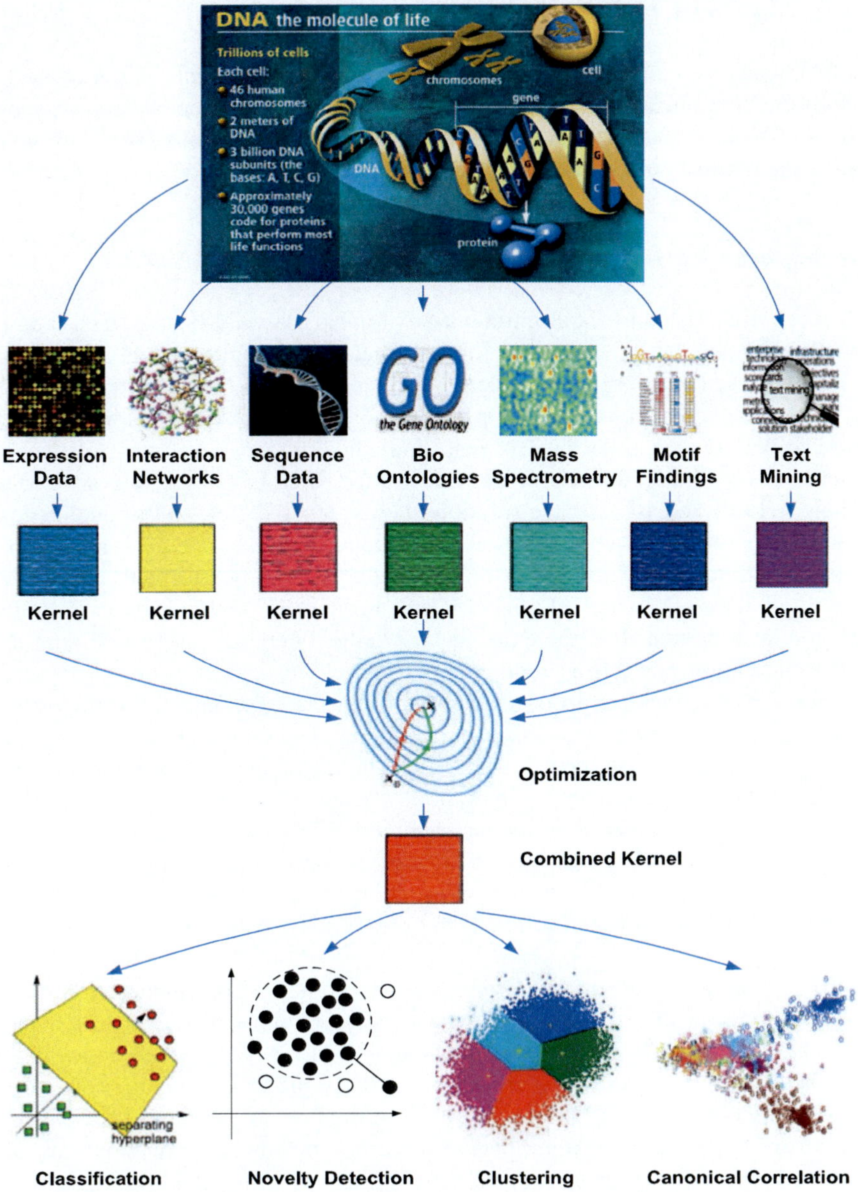


**Fig. 1.6** Conceptual map of kernel-based data fusion in Systems Biology. The DNA the molecule of life figure is reproduced from the genome programs of the U.S. Department of Energy Office of Science. The Gene Ontology icon adapted from the Gene Ontology Project. The text mining figure is used courtesy of Dashboard Insight (www.dashboardinsight.com). The optimization figure is taken from Wikimedia commons courtesy of the artist. The SVM classification figure is reproduced from the work of Looy *et al.* [49] with permission. The clustering figure is reproduced from the work of Cao [13] with permission.

## 1.3 Topics of This Book

In this book, we introduce several novel kernel fusion techniques in the context of supervised learning and unsupervised learning. At the same time, we apply the proposed techniques and algorithms to some real world applications. The main topics discussed in this book can be briefly highlighted as follows.

### *Non-sparse Kernel Fusion Optimized for Different Norms*

Current kernel fusion methods introduced by Lanckriet *et al.* [48] and Bach *et al.* [3] mostly optimize the $L_\infty$-norm of multiple kernels in the dual problem. This method is characterized as the sparse solution, which assigns dominant coefficients on one or two kernels. The sparse solution is useful to distinguish the relevant sources from irrelevant ones. However, in real biomedical applications, most of the data sources are well selected and processed, so they often have high relevance to the problem. In these cases, sparse solution may be too selective to thoroughly combine the complementary information in the data. In real biomedical applications, with a small number of sources that are believed to be truly informative, we would usually prefer a nonsparse set of coefficients because we would want to avoid that the dominant source (like the existing knowledge contained in Text Mining data and Gene Ontology) gets a dominant coefficient. The reason to avoid sparse coefficients is that there is a discrepancy between the experimental setup for performance evaluation and *real world* performance. The dominant source will work well on a benchmark because this is a controlled situation with known outcomes. In these cases, a sparse solution may be too selective to thoroughly combine the complementary information in the data sources. While the performance on benchmark data may be good, the selected sources may not be as strong on truly novel problems where the quality of the information is much lower. We may thus expect the performance of such solutions to degrade significantly on actual *real-world* applications.

To address this problem, we propose a new kernel fusion scheme to optimize the $L_2$-norm and the $L_n$-norm in the dual representations of kernel fusion models. The $L_2$-norm often leads to an non-sparse solution, which distributes the coefficients evenly on multiple kernels, and at the same time, leverages the effects of kernels in the objective optimization. Empirical results show that the $L_2$-norm kernel fusion may lead to better performance in biomedical applications. We also show that the strategy of optimizing different norms in the dual problem can be straightforwardly extended to any real number $n$ between 1 and 2, known as the $L_n$-norm kernel fusion. We found there is a simple mathematical relationship between the norm $m$ applied as the coefficient regularization in the primal problem with the norm $n$ of multiple kernels optimized in the dual problem. On this basis, we propose a set of convex solutions for the kernel fusion problem with arbitrary norms.

## *Kernel Fusion in Unsupervised Learning*

Kernel fusion is originally proposed for supervised learning and the problem is solved as a convex quadratic problem [9]. For unsupervised learning problem where the data samples are usually labeled or partially labeled, the optimization is often difficult and usually results in a non-convex solution where the global optimality is hard to determine. For example, the *k*-means clustering [7, 27] is solved as a non-convex stochastic process and it has lots of local minima. In this book, we present approaches to incorporate a non-convex unsupervised learning problem with the convex kernel fusion method, and the issues of convexity and convergence are tackled in an alternative minimization framework [20].

When kernel fusion is applied to unsupervised learning, the model selection problem becomes more challenging. For instance, in clustering problem the model evaluation usually relies on the statistical validation, which is often measured as various internal indices, such as Silhouette index [64], Jaccard index [41], Modularity [57], and so on. However, most of the internal indices are data dependent thus are not consistent with each other among heterogeneous data sources, which makes the model selection problem more difficult. In contrast, external indices evaluate models using the ground truth labels (e.g., Rand Index [39], Normalized Mutual Information [73]), which are more reliable to be used for optimal model selection. Unfortunately, the ground truth labels may not always be available for *real world* clustering problem. Therefore, how to select unsupervised learning model in data fusion applications is also one of the main challenges. In machine learning, most existing benchmark data sets are proposed for single source learning thus to validate data fusion approaches, people usually generate multiple data sources artificially using different distance measures on the same data set. In this way, the combined information is more likely to be redundant, which makes the approach less meaningful and less significant. Therefore, the true merit of data fusion should be demonstrated and evaluated in real applications using genuine heterogeneous data sources.

## *Kernel Fusion in Real Applications*

Kernel methods have been proved as powerful statistical learning techniques and they are widely applied to various learning scenarios due to their flexibility and good performance [60]. In recent years, many useful softwares and toolboxes of kernel methods have been developed. In particular, the kernel fusion toolbox is also recently proposed in Shogun software [71]. However, there is still a limit number of open source biomedical applications which are truly based on kernel methods or kernel fusion techniques. The gap between the algorithmic innovations and the real applications of kernel fusion methods is probably because of the following reasons. Firstly, the data preprocessing and data cleaning tasks in real applications often vary from problems to problems. Secondly, to tune the optimal kernel parameters and the hyper-parameters of the model on unseen data is a non-trivial task. Thirdly, most kernel fusion problems are solved by nonlinear optimization, which turns to be computational demanding when the data sets have very large scales.

In this book, we present a real bioinformatics software *MerKator*, whose main feature is the cross-species prioritization through kernel based genomic data fusion over multiple data sources and multiple species. To our knowledge, MerKator is one of the first real bioinformatics softwares powered by kernel methods. It is also one of the first cross-species prioritization softwares freely accessible online. To improve the efficiency of Merkator, we tackle the kernel computational challenges of full genomic data from multiple aspects. First, most of the kernels are pre-computed and preprocessed offline and performed only once, restricting the case specific online computation to a strict minimum. Second, the prioritization of the full genome utilizes some approximation techniques such as incomplete Cholesky decomposition, kernel centering in the subsets of genome, and missing value processing to improve its feasibility and efficiency.

## Large Scale Data and Computational Complexity

Unsupervised learning usually deals with large amount of data thus the computational burden of kernel fusion task is also large. In the supervised case, the model is often trained on a small number of labeled data and then generalized on the test data. Therefore, the main computational burden is determined by the training process whereas the complexity of model generalization on the test data is often linear. For example, given $N$ training data and $M$ test data, the computational complexity of the SVM training using a single kernel ranges from $O(N^2)$ to $O(N^3)$; the complexity of predicting labels on the test data is $O(M)$. In contrast, in unsupervised case one cannot split the data as training and test parts. The popular $k$-means clustering has the complexity of $O(k(N+M)dl)$, where $k$ is the number of clusters, $d$ is the complexity to compute the distance, and $l$ is the number of iterations. The kernel fusion procedure involving both training and test data has much larger computational burden than the supervised case. For instance, the *semi-definite programming* (SDP) solution of kernel fusion proposed by Lanckriet *et al.* [48] has the complexity up to $O((p+N+M)^2(k+N+M)^{2.5})$ [84]. When both $N$ and $M$ are big, kernel fusion is almost infeasible to be solved on a single node. This critical computational burden of kernel fusion can be tackled by various solutions from different aspects. In this book, we mainly focus on comparing various formulations of convex optimization and see how the selection of loss function in SVM could improve the efficiency of kernel fusion. Our main finding is, when the SVM objective is modeled on the basis of *Least squares support vector machines* (LSSVM) [76, 75] and the kernel fusion objective is modeled by *Semi-infinite programming* (SIP) [37, 42, 63, 70], the computational burden of kernel fusion can be significantly reduced as a limited iterations of linear problems. Of course, the efficiency of SVM kernel fusion can be further improved by various techniques, such as the active set method [14, 68], the gradient descent in the primal problem [61], the parallelization technique [70], and more recently the potential avenue explored in the Map/Reduce framework [24] for machine learning [15]. Fortunately, in a fast developing field, most of these approaches could be combined together to tackle the kernel fusion problem on very large scale dataset.

## 1.4 Chapter by Chapter Overview

*Chapter 2* investigates several unsupervised learning problems and summarizes their objectives as a common (generalized) *Rayleigh quotient* form. In particular, it shows the relationship between the Rayleigh quotient and the *Fisher Discriminant Analysis* (FDA), which serves as the basis of many machine learning methodologies. The FDA is also related to the kernel fusion approach formulated in least squares Support Vector Machines (LSSVM) [76, 75]. Clarifying this connection provides the theoretical grounding for us to incorporate kernel fusion methods in several concrete unsupervised algorithms.

*Chapter 3* extends kernel fusion, also known as *Multiple Kernel Learning* (MKL), to various machine learning problems. It proposes several novel results: Firstly, it generalizes the $L_\infty$ MKL formulation proposed by Lanckriet *et al.* and Bach *et al.* to a novel $L_2$ formulation, and further extends it to the arbitrary $L_n$-norm. The $L_\infty$-norm and $L_2$-norm differ at the norms optimized in terms of multiple kernels in the dual problem. Secondly, the chapter introduces the notion of MKL in LSSVM, which yields an efficient kernel fusion solution for large scale data. The connection between LSSVM MKL with FDA in the kernel space is also clarified, which serves as the core component in unsupervised algorithms and some relevant applications to be discussed in the remaining chapters.

*Chapter 4* extends kernel fusion to unsupervised learning and proposes a novel *Optimized kernel k-means Clustering* (OKKC) algorithm [91]. The algorithm tackles the non-convex optimization of multiple unlabeled data sources in a local alternative minimization framework [20]. The proposed algorithm is compared to some relevant work and its advantage is demonstrated as a simple objective and iterations of linear computations.

*Chapter 5* presents a real biomedical literature mining application using kernel fusion techniques of novelty detection and clustering proposed in Chapter 3 and Chapter 4. This approach combines several *Controlled Vocabularies* (CVs) using ensemble methods and kernel fusion methods to improve the accuracy of identifying disease relevant genes. Experimental result shows that the combination of multiple CVs in text mining can outperform the approaches using individual CVs alone. Thus it provides an interesting approach to exploit information combined by the myriad of different bio-ontologies.

*Chapter 6* proceeds the topic of Chapter 4 and considers the integration of kernel matrices with Laplacian matrices in clustering. We propose a novel algorithm, called *Optimized k-means Laplacian Clustering* (OKLC) [88], to combine the attribute representations based on kernels with the graph representation based on Laplacians in clustering analysis. Two real applications were investigated in this Chapter. The first one is improved from the literature mining results obtained from multiple CVs introduced in Chapter 5. Besides the relationship of disease relevant genes in terms of lexical similarities, we consider the spectral properties among them and combine the lexical similarities with spectral properties to further improve the accuracy of disease relevant clustering. In the second experiment, a Scientometrics application is demonstrated to combine attribute based lexical similarities with graph based citation links for journal mapping. The attribute information is transformed as kernels and the citations are represented as Laplacian matrices, then are all combined by OKLC to construct journal mapping by clustering. The merit of this approach is illustrated in a systematic evaluation with many comparing approaches and the proposed algorithm is shown outperforming over all other methods.

*Chapter 7* discusses Canonical Correlation Analysis, a different unsupervised learning problem than clustering. A new method called *Weighted Multiple Kernel Canonical Correlation Analysis* (WMKCCA) is proposed to leverage the importance of different data sources in the CCA objective [86]. Beside the derivation of mathematical models, we present some preliminary results of using the mappings obtained by WMKCCA as the common information extracted from multiple data sources.

*Chapter 8* continues to discuss the gene prioritization problem started in Chapter 5. To further exploits the information among genomic data sources and the phylogenetic evidences among different species, we design and develop an open software, MerKator [90], to perform cross-species gene prioritization by genomic data fusion. To our knowledge, it is one of the first real bioinformatics softwares powered by kernel fusion methods.

*Chapter 9* summarizes the book and highlights several topics that worth further investigation.

## References

1. Aizerman, M., Braverman, E., Rozonoer, L.: Theoretical foundations of the potential function method in pattern recognition learning. Automation and Remote Control 25, 821–837 (1964)
2. Bach, F.R., Jordan, M.I.: A Probabilistic Interpretation of Canonical Correlation Analysis. Internal Report 688, Department of Statistics. Department of Statistics, University of California, Berkeley (2005)
3. Bach, F.R., Jordan, M.I.: Kernel independent component analysis. Journal of Machine Learning Research 3, 1–48 (2003)
4. Baldwin, M.J.: A New Factor in Evolution. The American Naturalist 30, 441–451 (1896)
5. Barry, D.J.: Design Of and Studies With a Novel One Meter Multi-Element Spectroscopic Telescope. Ph.D dissertation, University of Cornell (1995)
6. Becker, S.: Mutual Information Maximization: models of cortical self-organization. Network: Computation in Neural System 7, 7–31 (1996)
7. Bishop, C.M.: Pattern recognition and machine learning. Springer, New York (2006)
8. Boser, B.E., Guyon, I.M., Vapnik, V.N.: A training algorithm for optimal margin classifiers. In: Proceedings of the 5th Annual ACM Workshop on COLT, pp. 144–152. ACM Press, New York (1992)
9. Boyd, S., Vandenberghe, L.: Convex Optimization. Cambridge University Press, Cambridge (2004)
10. Breiman, L.: Random forests. Machine Learning 45, 5–32 (2001)
11. Brieman, L.: Bagging predictors. Machine Learning 24, 123–140 (1996)
12. Breiman, L.: Statistical Modeling: The Two Cultures. Statistical Science 16, 199–231 (2001)

13. Cao, Y.: Efficient K-Means Clustering using JIT. MATLAB Central file exchange (2008), `http://www.mathworks.com/matlabcentral/fileexchange/19344-efficient-k-means-clustering-using-jit`
14. Cauwenberghs, G., Poggio, T.: Incremental and decremental support vector machine learning. Advances in Neural Information Processing Systems 13, 409–415 (2001)
15. Chu, C.T., Kim, S.K., Lin, Y.A., Yu, Y.Y., Bradski, G., Ng, A.Y., Olukotun, K.: Map-Reduce for Machine Learning on Multicore. Advances in Neural Information Processing Systems 20, 281–288 (2008)
16. Confucius: The Analects. 500 B.C
17. Cooper, G.F., Herskovits, E.: A Bayesian method for the induction of probabilistic networks from data. Machine Learning 9, 309–347 (1999)
18. Cover, T.M., Hart, P.E.: Nearest neighbor pattern classification. IEEE Trans. Information Theory. 13, 21–27 (1967)
19. Cristianini, N., Shawe-Taylor, J.: An Introduction to Support Vector Machines. Cambridge University Press, Cambridge (2000)
20. Csiszar, I., Tusnady, G.: Information geometry and alternating minimization procedures. Statistics and Decisions suppl. 1, 205–237 (1984)
21. Dash, D., Druzdzel, M.J.: Robust independence testing for constraint-based learning of causal structure. In: Proceedings of the 19th Conference on Uncertainty in Artificial Intelligence, pp. 167–174 (2003)
22. De Raedt, L., Dehaspe, L.: Clausal discovery. Machine Learning 26, 99–146 (1997)
23. De Raedt, L., Van Laer, W.: Inductive constraint logic. In: Zeugmann, T., Shinohara, T., Jantke, K.P. (eds.) ALT 1995. LNCS, vol. 997, pp. 80–94. Springer, Heidelberg (1995)
24. Dean, J., Ghemawat, S.: MapReduce: simplified data processing on large clusters. Communications of the ACM - 50th Anniversary issue: 1958 - 2008 51, 107–113 (2008)
25. Dempster, A.P., Laird, N.M., Rubin, D.B.: Maximum Likelihood from Incomplete Data via the EM Algorithm. Journal of the Royal Statistical Society. Series B (Methodological) 39, 1–38 (1977)
26. Drucker, H., Schapire, R., Simard, P.: Improving performance in neural networks using a boosting algorithm. Advances in Neural Information Processing Systems 5, 42–49 (1993)
27. Duda, R.O., Hart, P.E., Stork, D.G.: Pattern Classification, 2nd edn. John Wiley & Sons Inc., New York (2001)
28. Evans, J., Rzhetsky, A.: Machine Science. Science 329, 399–400 (2010)
29. Freund, Y., Schapire, R.: A decision-theoretic generalization of online learning and an application to boosting. Journal of Computer and System Sciences 55, 119–139 (1997)
30. Friedman, N.: Learning belief networks in the presence of missing values and hidden variables. In: Proceedings of the 14th ICML, pp. 125–133 (1997)
31. Friedman, C., Borlawsky, T., Shagina, L., Xing, H.R., Lussier, Y.A.: Bio-Ontology and text: bridging the modeling gap. Bioinformatics 22, 2421–2429 (2006)
32. Fromont, E., Quiniou, R., Cordier, M.-O.: Learning Rules from Multisource Data for Cardiac Monitoring. In: Miksch, S., Hunter, J., Keravnou, E.T. (eds.) AIME 2005. LNCS (LNAI), vol. 3581, pp. 484–493. Springer, Heidelberg (2005)
33. Galperin, M.Y.: The Molecular Biology Database Collection: 2008 Update. Nucleic acids research 4, D2–D4 (2008)
34. Gevaert, O.: A Bayesian network integration framework for modeling biomedical data. Ph.D dissertation, Katholieke Universiteit Leuven (2008)
35. Hardoon, D.R., Shawe-Taylor, J.: Canonical Correlation Analysis: An Overview with Application to Learning Methods. Neural Computation 16, 2639–2664 (2004)

36. Hastie, T., Tibshirani, R., Friedman, J.: The Elements of Statistical Learning: Data Mining, Inference, and Prediction, 2nd edn. Springer, Heidelberg (2009)
37. Hettich, R., Kortanek, K.O.: Semi-infinite programming: theory, methods, and applications. SIAM Review 35, 380–429 (1993)
38. Hotelling, H.: Relations between two sets of variates. Biometrika 28, 321–377 (1936)
39. Hubert, L., Arabie, P.: Comparing partitions. Journal of Classification 2, 193–218 (1985)
40. Hucka, M., Finney, A., Sauro, H.M., et al.: The systems biology markup language (SBML): a medium for representation and exchange of biochemical network models. Bioinformatics 19, 524–531 (2003)
41. Jaccard, P.: Distribution de la flore alpine dans le bassin des Dranses et dans quelques régions voisines. Bulletin de la Société Vaudoise des Sciences Naturelles 37, 241–272 (1901)
42. Kaliski, J., Haglin, D., Roos, C., Terlaky, T.: Logarithmic barrier decomposition methods for semi-infinite programming. International Transactions in Operations Research 4, 285–303 (1997)
43. Klami, A., Kaski, S.: Generative models that discover dependencies between two data sets. In: Proc. of IEEE Machine Learning for Signal Processing XVI, pp. 123–128 (2006)
44. Kloft, M., Brefeld, U., Laskov, P., Sonnenburg, S.: Non-sparse Multiple Kernel Learning. In: NIPS 2008 Workshop: Kernel Learning - Automatic Selection of Optimal Kernels (2008)
45. Krogh, A., Vedelsby, J.: Neural network ensembles, cross-validation and active learning. Advances in Neural Information Processing Systems 7, 231–238 (1995)
46. Lai, P.L., Fyfe, C.: Kernel and Nonlinear Canonical Correlation Analysis. International Journal of Neural Systems 10, 365–377 (2000)
47. Lanckriet, G.R.G., Cristianini, N., Jordan, M.I., Noble, W.S.: Kernel Methods in Computational Biology. MIT Press, Cambridge (2004)
48. Lanckriet, G.R.G., De Bie, T., Cristianini, N., Jordan, M.I., Noble, W.S.: A statistical framework for genomic data fusion. Bioinformatics 20, 2626–2635 (2004)
49. Looy, S.V., Verplancke, T., Benoit, D., Hoste, E., Van Maele, G., De Turck, F., Decruyenaere, J.: A novel approach for prediction of tacrolimus blood concentration in liver transplantation patients in the intensive care unit through support vector regression. Critical Care 11, R83 (2007)
50. Lloyd, J.: Foundations of Logic Programming. Springer, New York (1987)
51. Mika, S., Rätsch, G., Weston, J., Schölkopf, B.: Fisher discriminant analysis with kernels. In: IEEE Neural Networks for Signal Processing IX: Proceedings of the 1999 IEEE Signal Processing Society Workshop, pp. 41–48 (1999)
52. Mika, S., Weston, J., Schölkopf, B., Smola, A., Müller, K.-R.: Constructing Descriptive and Discriminative Nonlinear Features: Rayleigh Coefficients in Kernel Feature Spaces. IEEE Trans. on PAMI 25, 623–628 (2003)
53. Muggleton, S., De Raedt, L.: Inductive Logic Programming: Theory and methods. The Journal of Logic Programming 19/20, 629–680 (1994)
54. Myers, J.W.: Learning bayesian network from incomplete data with stochastic search algorithms. In: Proceedings of the 15th Conference on Uncertainty in Artificial Intelligence, pp. 476–485. Morgan Kaufmann Publishers, San Francisco (1999)
55. Needham, C.J., Bradford, J.R., Bulpitt, A.J., Westhead, D.R.: A Primer on Learning in Bayesian Networks for Computational Biology. PLOS Computational Biology 3, 1409–1416 (2007)
56. Nesterov, Y., Nemirovskij, A.: Interior-point polynomial algorithms in convex programming. SIAM Press, Philadelphia (1994)

57. Newman, M.E.J.: Modularity and community structure in networks. Proc. Natl. Acad. Sci. USA. 103, 8577–8582 (2006)
58. Parzen, E.: On Estimation of a Probability Density Function and Mode. Annals of Mathematical Statistics 33, 1065–1076 (1962)
59. Pearl, J.: Probabilistic Reasoning in Intelligent Systems: Networks of Plausible Inference. Morgan Kaufmann Publishers, San Francisco (1988)
60. Pekalska, E., Haasdonk, B.: Kernel Discriminant Analysis for Positive Definite and Indefinite Kernels. IEEE Trans. TPAMI 31, 1017–1031 (2009)
61. Rakotomamonjy, A., Bach, F.R., Canu, S., Grandvalet, Y.: Simple MKL. Journal of Machine Learning Research 9, 2491–2521 (2008)
62. Ramoni, M., Sebastiani, P.: Robust learning with missing data. Machine Learning 45, 147–170 (2000)
63. Reemtsen, R.: Some other approximation methods for semi-infinite optimization problems. Jounral of Computational and Applied Mathematics 53, 87–108 (1994)
64. Rousseeuw, P.J.: Silhouettes: a graphical aid to the interpretation and validation of cluster analysis. Journal of Computational and Applied Mathematics 20, 53–65 (1987)
65. Santosh, K.C., Lamiroy, B., Ropers, J.-P.: Inductive Logic Programming for Symbol Recognition. In: Proc. of the 10th International Conference on Document Analysis and Recognition, pp. 1330–1334 (2009)
66. Schölkopf, B., Smola, A.: Learning with Kernels. MIT Press, Cambridge (2002)
67. Schölkopf, B., Smola, A., Müller, K.-R.: Nonlinear component analysis as a kernel eigenvalue problem. Neural Computation 10, 1299–1319 (1998)
68. Scheinberg, K.: An Efficient Implementation of an Active Set Method for SVMs. Journal of Machine Learning Research 7, 2237–2257 (2006)
69. Shapior, B.E., Hucka, M., Finney, A., Doyle, J.: MathSBML: a package for manipulating SBML-based biological models. Bioinformatics 20, 2829–2831 (2004)
70. Sonnenburg, S., Rätsch, G., Schäfer, C., Schölkopf, B.: Large Scale Multiple Kernel Learning. Jounral of Machine Learning Research 7, 1531–1565 (2006)
71. Sonnenburg, S., Räetsch, G., Henschel, S., Widmer, C., Behr, J., Zien, A., de Bona, F., Binder, A., Gehl, C., Franc, V.: The SHOGUN Machine Learning Toolbox. Journal of Machine Learning Research 11, 1799–1802 (2010)
72. Stephen, W.: Primal-Dual Interior-Point Methods. SIAM Press, Philadelphia (1997)
73. Strehl, A., Ghosh, J.: Cluster Ensembles: A Knowledge Reuse Framework for Combining Multiple Partitions. Journal of Machine Learning Research 3, 583–617 (2002)
74. Sutton, C.D.: Classification and Regression Trees, Bagging, and Boosting. Handbook of Statistics 24, 303–329 (2005)
75. Suykens, J.A.K., Van Gestel, T., De Brabanter, J., De Moor, B., Vandewalle, J.: Least Squares Support Vector Machines. World Scientific Press, Singapore (2002)
76. Suykens, J.A.K., Vandewalle, J.: Least squares support vector machine classifiers. Neural Processing Letters 9, 293–300 (1999)
77. Taton, R.: La premire note mathmatique de Gaspard Monge (juin 1769). Rev. Histoire Sci. Appl. 19, 143–149 (1966)
78. Taylor, C.F., Paton, N.W., Garwood, K.L., et al.: A systematic approach to modeling, capturing, and disseminating proteomics experimental data. Nature Biotechnology 21, 247–254 (2003)
79. Tretyakov, K.: Methods of Genomic Data Fusion: An Overview. Technical Report, Institute of Computer Science, University of Tartu (2006)
80. Vapnik, V.: The Nature of Statistical Learning Theory, 2nd edn. Springer, New York (1999)

81. Vapnik, V.: Statistical Learning Theory. Wiley Interscience, New York (1998)
82. Vapnik, V., Chervonenkis, A.: On the uniform convergence of relative frequencies of events to their probabilities. Theory of Probability and its Applications 16, 264–280 (1971)
83. Wikipedia: Antikythera mechanism, `http://en.wikipedia.org/wiki/Antikythera_mechanism`
84. Ye, J.P., Ji, S.W., Chen, J.H.: Multi-class Discriminant Kernel Learning via Convex Programming. Jounral of Machine Learning Research 9, 719–758 (2008)
85. Yu, K., Ji, L., Zhang, X.G.: Kernel Nearest-Neighbor Algorithm. Neural Processing Letters 15, 147–156 (2002)
86. Yu, S., De Moor, B., Moreau, Y.: Learning with heterogeneous data sets by Weighted Multiple Kernel Canonical Correlation Analysis. In: Proc. of the Machine Learning for Signal Processing XVII, pp. 81–86. IEEE, Los Alamitos (2007)
87. Yu, S., Falck, T., Tranchevent, L.-C., Daemen, A., Suykens, J.A.K., De Moor, B., Moreau, Y.: L2-norm multiple kernel learning and its application to biomedical data fusion. BMC Bioinformatics 11, 1–53 (2010)
88. Yu, S., Liu, X.H., Glänzel, W., De Moor, B., Moreau, Y.: Optimized data fusion for K-means Laplacian Clustering. Bioinformatics 26, 1–9 (2010)
89. Yu, S., Tranchevent, L.-C., De Moor, B., Moreau, Y.: Gene prioritization and clustering by multi-view text mining. BMC Bioinformatics 11, 1–48 (2010)
90. Yu, S., Tranchevent, L.-C., Leach, S., De Moor, B., Moreau, Y.: Cross-species gene prioritization by genomic data fusion. Internal Report (2010) (submitted for publication)
91. Yu, S., Tranchevent, L.-C., Liu, X., Glänzel, W., Suykens, J.A.K., De Moor, B., Moreau, Y.: Optimized data fusion for kernel K-means clustering. Internal Report 08-200, ESAT-SISTA, K.U.Leuven, Lirias number: 242275 (2008) (submitted for publication)
92. Zheng, W.J.: Engineering Approaches Toward Biological Information Integration at the Systems Level. Current Bioinformatics 1, 85–93 (2006)

# Chapter 2
# Rayleigh Quotient-Type Problems in Machine Learning

## 2.1 Optimization of Rayleigh Quotient

### 2.1.1 *Rayleigh Quotient and Its Optimization*

For real matrices and vectors, given a positive definite matrix $Q$ and a nonzero vector $\mathbf{w}$, a Rayleigh quotient (also known as *Rayleigh-Ritz ratio*) is defined as

$$\rho = \rho(\mathbf{w}; Q) = \frac{\mathbf{w}^T Q \mathbf{w}}{\mathbf{w}^T \mathbf{w}}. \tag{2.1}$$

It was originally proposed in the theorem to approximate eigenvalues of a Hermitian matrix. The theorem is mentioned as as follows:

**Theorem 2.1.** *(Rayleigh-Ritz) Let $Q \in M_n$ be Hermitian, and let the eigenvalues of $Q$ be ordered as $\lambda_{min} = \lambda_1 \leq \lambda_2 \leq \cdots \leq \lambda_{n-1} \leq \lambda_n = \lambda_{max}$. Then*

$$\lambda_1 \mathbf{w}^T \mathbf{w} \leq \mathbf{w}^T Q \mathbf{w} \leq \lambda_n \mathbf{w}^T \mathbf{w} \ \textit{for all} \ \mathbf{w} \in \mathbb{C}^n,$$

$$\lambda_{max} = \lambda_n = \max_{\mathbf{w} \neq 0} \frac{\mathbf{w}^T Q \mathbf{w}}{\mathbf{w}^T \mathbf{w}} = \max_{\mathbf{w}^T \mathbf{w} = 1} \mathbf{w}^T Q \mathbf{w},$$

$$\lambda_{min} = \lambda_1 = \min_{\mathbf{w} \neq 0} \frac{\mathbf{w}^T Q \mathbf{w}}{\mathbf{w}^T \mathbf{w}} = \min_{\mathbf{w}^T \mathbf{w} = 1} \mathbf{w}^T Q \mathbf{w}.$$

*Proof.* See Theorem 4.2.2 in [4].

In machine learning, many problems can be simplified as the minimization or maximization of the Rayleigh quotient, given by

$$\underset{\mathbf{w}}{\text{maximize}} \frac{\mathbf{w}^T Q \mathbf{w}}{\mathbf{w}^T \mathbf{w}}, \quad \text{or} \ \underset{\mathbf{w}}{\text{minimize}} \frac{\mathbf{w}^T Q \mathbf{w}}{\mathbf{w}^T \mathbf{w}}. \tag{2.2}$$

Notice that the quotient is invariant to the magnitude of $\mathbf{w}$, one can reformulate the problem in terms of quadratic programming, given by

S. Yu et al.: Kernel-based Data Fusion for Machine Learning, SCI 345, pp. 27–37.
springerlink.com 

$$\begin{aligned} &\underset{\mathbf{w}}{\text{maximize}} \;\; \mathbf{w}^T Q \mathbf{w} \\ &\text{subject to} \;\; \mathbf{w}^T \mathbf{w} = 1. \end{aligned} \tag{2.3}$$

The Lagrangian $\mathscr{L}(\mathbf{w}, \lambda)$ is

$$\mathscr{L}(\mathbf{w}, \lambda) = \underset{\mathbf{w},\lambda}{\text{maximize}} \left( \mathbf{w}^T Q \mathbf{w} - \lambda (\mathbf{w}^T \mathbf{w} - 1) \right), \tag{2.4}$$

where $\lambda \in \mathbb{R}$ is the Lagrangian multipliers (dual variable). Taking the conditions for optimality from the Lagrangian $\partial \mathscr{L} / \partial \mathbf{w} = 0$, $\partial \mathscr{L} / \partial \lambda = 0$, (2.4) leads to the eigenvalue problem as

$$Q\mathbf{w} = \lambda \mathbf{w}. \tag{2.5}$$

Obviously, the optimal solution of the maximization problem in (2.2) is obtained by $\mathbf{w}_{max}$ in the maximal eigenvalue pair $(\lambda_{max}, \mathbf{w}_{max})$ of $Q$. Similarly, the solution of minimization is corresponding to $\mathbf{w}_{min}$ in the minimal eigenvalue pair $(\lambda_{min}, \mathbf{w}_{min})$.

### 2.1.2 Generalized Rayleigh Quotient

The Generalized Rayleigh quotient involves two quadratic forms, given by

$$\rho(\mathbf{w}; Q, P) = \frac{\mathbf{w}^T Q \mathbf{w}}{\mathbf{w}^T P \mathbf{w}}, \tag{2.6}$$

where $Q$ and $P$ are positive definite matrices, $\mathbf{w} \neq 0$. The problem is to maximize or minimize the Generalized Rayleigh quotient as follows:

$$\underset{\mathbf{w}}{\text{maximize}} \frac{\mathbf{w}^T Q \mathbf{w}}{\mathbf{w}^T P \mathbf{w}}, \quad \text{or} \quad \underset{\mathbf{w}}{\text{minimize}} \frac{\mathbf{w}^T Q \mathbf{w}}{\mathbf{w}^T P \mathbf{w}}. \tag{2.7}$$

Taking the conditions for optimality from the Lagrangian, one obtains

$$Q\mathbf{w} = \lambda P \mathbf{w}. \tag{2.8}$$

Thus the optimum of (2.7) can be obtained by exploiting the eigenvalue pair of the generalized eigenvalue problem in (2.8).

### 2.1.3 Trace Optimization of Generalized Rayleigh Quotient-Type Problems

We now consider a more general problem than (2.7) by replacing $\mathbf{w}$ with a matrix $W = [\mathbf{w}_1, ..., \mathbf{w}_k]$. Firstly, we quote the Ky-Fan theorem as

**Theorem 2.2.** *(Ky Fan) Let $Q \in \mathbb{R}^{n \times n}$ be a positive definite matrix, and let the eigenvalues of $Q$ be ordered as $\lambda_{min} = \lambda_1 \leq \lambda_2 \leq \ldots \leq \lambda_{n-1} \leq \lambda_n = \lambda_{max}$ and the corresponding eigenvectors as $\mathbf{u}_1, ..., \mathbf{u}_n$. Then*

$$\lambda_1 + \cdots + \lambda_k = \underset{W^T W = I}{\text{maximize}} \ \text{trace}(W^T Q W). \tag{2.9}$$

*Moreover, the optimal $W^*$ is given by $W^* = [\mathbf{u}_1, ..., \mathbf{u}_k]U$, where $U$ is an arbitrary orthogonal matrix.*

*Proof.* See [1, 12].

To consider two positive definite matrices $Q$ and $P$, one generalizes the Ky-Fan theorem as follows:

**Theorem 2.3.** *Given a real value matrix $W = [\mathbf{w}_1, ..., \mathbf{w}_k] \in \mathbb{R}^{n \times k}$, and two positive definite matrices $Q, P \in \mathbb{R}^{n \times n}$, the optimal solution of the trace maximization problem as*

$$\begin{aligned} & \underset{W}{\text{maximize}} \ \text{trace}(W^T Q W)^{-1}(W^T P W) \\ & \text{subject to} \ \ W^T W = I, \end{aligned} \tag{2.10}$$

*is obtained by the k vectors corresponding to the largest k eigenvalue pairs solved from the generalized eigenvalue problem*

$$Q\xi = \lambda P \xi. \tag{2.11}$$

*Proof.* Denote

$$F = P^{\frac{1}{2}} W, \tag{2.12}$$

then

$$F^T F = W^T P W. \tag{2.13}$$

Denote

$$P = U \Lambda U^T \tag{2.14}$$

as the eigenvalue decomposition of $P$ where $\Lambda$ is the diagonal matrix of the eigenvalues, given by

$$(F^T F)^{-\frac{1}{2}} = U \Lambda^{-\frac{1}{2}} U^T. \tag{2.15}$$

Since $W^T P W$ is symmetric, the quotient term in (2.10) can be equivalently rewritten as

$$\begin{aligned} (W^T P W)^{-1}(W^T Q W) &= (W^T P W)^{-\frac{1}{2}}(W^T Q W)(W^T P W)^{-\frac{1}{2}} \\ &= (F^T F)^{-\frac{1}{2}}(W^T Q W)(F^T F)^{-\frac{1}{2}} \\ &= (F^T F)^{-\frac{1}{2}} F^T P^{-\frac{1}{2}} Q P^{-\frac{1}{2}} F (F^T F)^{-\frac{1}{2}}. \end{aligned} \tag{2.16}$$

Let us denote that $B = F(F^T F)^{-\frac{1}{2}}$ and replace (2.15) in (2.16), moreover, since $B^T B = I$, thus (2.16) is equal to

$$(W^T PW)^{-1}(W^T QW) = B^T P^{-\frac{1}{2}} QP^{-\frac{1}{2}} B. \tag{2.17}$$

According to the Ky-Fan, the optimal of maximizing (2.17) is given by the eigenvectors corresponding to the largest eigenvalues of $P^{-\frac{1}{2}} QP^{-\frac{1}{2}}$, which is equivalent to the solution of the generalized eigenvalue problem in (2.9). □

The formal proof of Theorem 2.3 is given in [6, 10].

We have now introduced the basic formulations of Rayleigh quotient-type problems and their optimal solutions. Next, we will show that many unsupervised learning problems can be simplified to the Rayleigh quotient-type forms.

## 2.2 Rayleigh Quotient-Type Problems in Machine Learning

### 2.2.1 *Principal Component Analysis*

*Principal Component Analysis* (PCA) considers a given set of zero mean data $X \in \mathbb{R}^{N\times d}$, where $X = \{\mathbf{x}_i\}_{i=1}^N$, $\mathbf{x}_i \in \mathbb{R}^d$. The objective is to find a direction vector $\mathbf{w}$ on which the variance of the projection $\mathbf{w}^T \mathbf{x}_i$ is maximized. Since the variance is invariant to the magnitude of $\mathbf{w}$, the objective of PCA is equivalently formulated as

$$\begin{aligned} &\underset{\mathbf{w}}{\text{maximize}} \quad \mathbf{w}^T C_{\mathbf{xx}} \mathbf{w} \\ &\text{subject to} \quad \mathbf{w}^T \mathbf{w} = 1, \end{aligned} \tag{2.18}$$

where $C_{\mathbf{xx}}$ is the sample covariance matrix of $X$. Obviously, (2.18) is a Rayleigh quotient optimization and the optimal $\mathbf{w}$ is given by the eigenvector from the largest eigenvalue pair of $C_{\mathbf{xx}}$.

### 2.2.2 *Canonical Correlation Analysis*

*Canonical Correlation Analysis* (CCA) finds linear relations between two sets of variables [5]. For two zero mean data sets $X \in \mathbb{R}^{N\times d_1}$ and $Y \in \mathbb{R}^{N\times d_2}$, the objective is to identify vectors $\mathbf{w}_1$ and $\mathbf{w}_2$ such that the correlation between the projected variables $\mathbf{w}_1^T X$ and $\mathbf{w}_2^T Y$ is maximized, given by

$$\underset{\mathbf{w}_1,\mathbf{w}_2}{\text{maximize}} \frac{\mathbf{w}_1^T C_{\mathbf{xy}} \mathbf{w}_2}{\sqrt{\mathbf{w}_1^T C_{\mathbf{xx}} \mathbf{w}_1} \sqrt{\mathbf{w}_2^T C_{\mathbf{yy}} \mathbf{w}_2}}, \tag{2.19}$$

where $C_{\mathbf{xx}} = \mathscr{E}[XX^T], C_{\mathbf{yy}} = \mathscr{E}[YY^T], C_{\mathbf{xy}} = \mathscr{E}[XY^T]$. The problem in (2.19) is usually formulated as the optimization problem, given by

$$\begin{aligned}\underset{\mathbf{w}_1,\mathbf{w}_2}{\text{maximize}} \quad & \mathbf{w}_1^T C_{\mathbf{xy}} \mathbf{w}_2 \\ \text{subject to} \quad & \mathbf{w}_1^T C_{\mathbf{xx}} \mathbf{w}_1 = 1, \\ & \mathbf{w}_2^T C_{\mathbf{yy}} \mathbf{w}_2 = 1.\end{aligned} \tag{2.20}$$

Taking the conditions for optimality from the Lagrangian

$$\mathscr{L}(\mathbf{w}_1,\mathbf{w}_2;\lambda_1,\lambda_2) = \mathbf{w}_1^T C_{\mathbf{xy}} \mathbf{w}_2 - \lambda_1(\mathbf{w}_1^T C_{\mathbf{xx}} \mathbf{w}_1 - 1) - \lambda_2(\mathbf{w}_2^T C_{\mathbf{yy}} \mathbf{w}_2 - 1), \tag{2.21}$$

one has

$$\begin{cases} C_{\mathbf{xy}} \mathbf{w}_2 = \lambda_1 C_{\mathbf{xx}} \mathbf{w}_1 \\ C_{\mathbf{yx}} \mathbf{w}_1 = \lambda_2 C_{\mathbf{yy}} \mathbf{w}_2 \end{cases}. \tag{2.22}$$

Since we have

$$\begin{cases} \mathbf{w}_1^T C_{\mathbf{xy}} \mathbf{w}_2 = \lambda_1 \mathbf{w}_1^T C_{\mathbf{xx}} \mathbf{w}_1 \\ \mathbf{w}_2^T C_{\mathbf{yx}} \mathbf{w}_1 = \lambda_2 \mathbf{w}_2^T C_{\mathbf{yy}} \mathbf{w}_2 \\ \mathbf{w}_1^T C_{\mathbf{xy}} \mathbf{w}_2 = \mathbf{w}_2^T C_{\mathbf{yx}} \mathbf{w}_1 \\ \mathbf{w}_1^T C_{\mathbf{xx}} \mathbf{w}_1 = \mathbf{w}_2^T C_{\mathbf{yy}} \mathbf{w}_2 \end{cases}, \tag{2.23}$$

we find that $\lambda_1 = \lambda_2 = \lambda$, thus we obtain a generalized eigenvalue problem, given by

$$\begin{bmatrix} 0 & C_{\mathbf{xy}} \\ C_{\mathbf{yx}} & 0 \end{bmatrix} \begin{bmatrix} \mathbf{w}_1 \\ \mathbf{w}_2 \end{bmatrix} = \lambda \begin{bmatrix} C_{\mathbf{xx}} & 0 \\ 0 & C_{\mathbf{yy}} \end{bmatrix} \begin{bmatrix} \mathbf{w}_1 \\ \mathbf{w}_2 \end{bmatrix}. \tag{2.24}$$

Analogously, the objective function of CCA can be also rewritten in a generalized Rayleigh quotient form as

$$\underset{\mathbf{w}_1,\mathbf{w}_2}{\text{maximize}} \frac{\begin{bmatrix} \mathbf{w}_1 \\ \mathbf{w}_2 \end{bmatrix}^T \begin{bmatrix} 0 & C_{\mathbf{xy}} \\ C_{\mathbf{yx}} & 0 \end{bmatrix} \begin{bmatrix} \mathbf{w}_1 \\ \mathbf{w}_2 \end{bmatrix}}{\begin{bmatrix} \mathbf{w}_1 \\ \mathbf{w}_2 \end{bmatrix}^T \begin{bmatrix} C_{\mathbf{xx}} & 0 \\ 0 & C_{\mathbf{yy}} \end{bmatrix} \begin{bmatrix} \mathbf{w}_1 \\ \mathbf{w}_2 \end{bmatrix}}. \tag{2.25}$$

### 2.2.3 *Fisher Discriminant Analysis*

*Fisher Discriminant Analysis* (FDA) optimizes the discriminating direction for classification, which is also expressed as a form similar to the Generalized Rayleigh quotient, where $S_B$ is the measure of the separability of class (between class scatter) and $S_W$ is the measure of within class scatter, given by

$$\underset{\mathbf{w}}{\text{maximize}} \frac{\mathbf{w}^T S_B \mathbf{w}}{\mathbf{w}^T S_W \mathbf{w}}, \tag{2.26}$$

where

$$S_B = (\mu_2 - \mu_1)(\mu_2 - \mu_1)^T,$$
$$S_W = \sum_{i=1,2} \sum_{\mathbf{x} \in \mathscr{X}_i} (\mathbf{x} - \mu_i)(\mathbf{x} - \mu_i)^T,$$

and $\mu_i$ denotes the sample mean for class $i$ [9].

### 2.2.4 *k-means Clustering*

It has been shown that (*e.g.*, [3]) the principal components in PCA are equivalent to the continuous solutions of the cluster membership indicators in the $k$-means clustering method. $k$-means uses $k$ number of prototypes to characterize the data and the partitions are determined by minimizing the variance

$$J_{k-\text{means}} = \sum_{a=1}^{k} \sum_{i \in \mathscr{X}_i} (\mathbf{x}_i - \mu_a)(\mathbf{x}_i - \mu_a)^T. \tag{2.27}$$

For a given data set $X$ and a cluster number $k$, the summation of all the pairwise distances is a constant value hence minimizing the distortion is equivalent to maximizing the *between clusters variance*, given by

$$J_{k-\text{means}} = \sum_{a=1}^{k} (\mu_a - \hat{\mu})(\mu_a - \hat{\mu})^T. \tag{2.28}$$

where $\hat{\mu}$ is the global sample mean of $X$.

Denote $P$ as the weighted cluster indicator matrix for $k$ classes, given by

$$A = F(F^T F)^{-\frac{1}{2}}, \tag{2.29}$$

where $F$ is the $N \times k$ binary cluster indicator matrix as

$$F = f_{i,j_{N\times k}}, \text{where } f_{i,j} = \begin{cases} 1 & \text{if } x_i \in l_j \\ 0 & \text{if } x_i \notin l_j \end{cases}. \tag{2.30}$$

Assume $X$ has zero mean, without losing generality, (2.28) can be re-written in the matrix form as

$$J_{k-\text{means}} = \underset{A}{\text{maximize}} \ \text{trace}\left(A^T X^T X A\right). \tag{2.31}$$

Because the construction of $A$ in (2.29) and (2.30) ensures that $A^T A = I$, the objective of $k$-means is exactly the maximization of a Rayleigh quotient. When $k = 2$, $A$ reduces to vector $\mathbf{a}$, and leads to a PCA problem. When $k > 2$, the cluster indicators $F$ can be recovered by exploiting the $k - 1$ principal components of $X^T X$, for instance, by QR decomposition proposed in [15]. This PCA based approach of $k$-means clustering is also known as *the spectral relaxation* of $k$-means.

### 2.2.5 Spectral Clustering

Spectral clustering models the data as graphs where the data samples are represented as vertices connected by non-negative weighted undirected edges. The clustering problem is then restated as to find a partition of the graph that the edges between different groups have a very low weight [7]. Different criteria have been applied to model the objective of cut for example, the RatioCut [2], the normalized cut [11], Markov Random Walks [8], the min-cut [13] and so on. In this book, the discussions about spectral clustering are all based on the normalized cut objective.

Let us denote $G = (V,E)$ as an undirected graph with vertex set $V = \{v_1,...,v_n\}$, $W$ as the weighted adjacency matrix of the graph $W = \{w_{ij}\}_{i,j=1,...,n}$, $d_i = \sum_{j=1}^{n} w_{ij}$ as the degree of a vertex $v_i \in V$, and $D$ as the diagonal matrix with the degrees $d_1,...,d_n$ on the diagonal. Given a subset of vertices $\mathscr{X} \subset V$, we denote its complement $V \setminus \mathscr{X}$ as $\bar{\mathscr{X}}$. For two disjoint subsets $M,N \subset V$, the cut is defined as

$$\mathrm{cut}(M,N) = \sum_{i \in M, j \in N} w_{ij} \ . \tag{2.32}$$

The size of a subset is defined as

$$\mathrm{vol}(M) = \sum_{i \in M} d_i. \tag{2.33}$$

The normalized cut criterion optimizes the partition $\mathscr{X}_1,\ldots,\mathscr{X}_k$ to minimize the objective as

$$\mathrm{Ncut}(\mathscr{X}_1,\ldots,\mathscr{X}_k) = \sum_{i=1}^{k} \frac{\mathrm{cut}(\mathscr{X}_i, \bar{\mathscr{X}_i})}{\mathrm{vol}(\mathscr{X}_i)}. \tag{2.34}$$

Unfortunately, to obtain the exact solution of (2.34) is NP hard [13]. To solve it, the discrete constraint of clustering indicators is usually relaxed to real values thus the approximated solution of spectral clustering can be obtained from the eigenspectrum of the graph Laplacian matrix. For $k$-way clustering ($k > 2$), the weighted cluster indicator matrix $P$ is defined in the same way as (2.29) and (2.30), the problem of minimizing the normalized cut is equivalently expressed as

$$\begin{aligned} &\underset{A}{\text{minimize}} \ \ \mathrm{trace}(A^T D^{-\frac{1}{2}} L D^{-\frac{1}{2}} A), \\ &\text{subject to} \ \ A^T A = I. \end{aligned} \tag{2.35}$$

This is again the optimization of a Rayleigh quotient problem which can be solved by eigenvalue decomposition. The optimal $A^*$ corresponds to the first $k$ eigenvectors of the normalized Laplacian $\tilde{L} = D^{-\frac{1}{2}} L D^{-\frac{1}{2}}$.

### 2.2.6 Kernel-Laplacian Clustering

Let us assume that the attribute based data $X$ and the graph affinity matrix $W$ are representations of the same sets of samples, the objective function of *Kernel-Laplacian* (KL) clustering can be defined as

$$J_{KL} = \kappa J_{Ncut} + (1-\kappa) J_{k-\text{means}} \tag{2.36}$$

where $\kappa$ is the weight adjusting the effect of $k$-means and spectral clustering objectives. $A$ is the weighted cluster indicator matrix as defined before. Replace (2.31) and (2.35) in (2.36), the objective of KL clustering becomes

$$\begin{aligned} J_{KL} &= \kappa \min_A \text{trace}\left(A^T \tilde{L} A\right) + (1-\kappa) \max_A \text{trace}\left(A^T X^T X A\right) \\ \text{s.t. } & A^T A = I. \end{aligned} \tag{2.37}$$

To solve the optimization problem without tuning the hyperparameter $\kappa$, Wang *et al.* propose a solution to optimize the trace quotient of the two sub-objectives [14]. The trace quotient formulation is then further relaxed as a maximization of the quotient trace, given by

$$\begin{aligned} J_{KL} &= \underset{A}{\text{maximize}} \;\; \text{trace}\left\{(A^T \tilde{L} A)^{-1} (A^T X^T X A)\right\} \\ & \text{subject to } \; A^T A = I. \end{aligned} \tag{2.38}$$

The objective in (2.38) is again a generalized Rayleigh quotient problem and the optimal solution $A^*$ is obtained by solving the generalized eigenvalue problem. To maximize the objective with $k$ clusters, $A^*$ is approximated as the largest $k$ eigenvectors of $\tilde{L}^+ \left(X^T X\right)$, where $\tilde{L}^+$ is the pseudo inverse of $\tilde{L}$ [14].

### 2.2.7 One Class Support Vector Machine

The *One class support vector machine* (1-SVM) method transforms the binary SVM classification task as one class learning problem. The method transforms the training data of one class into a high dimensional Hilbert space by the feature map, and iteratively finds the maximal margin hyper-plane that best separates the training data from the origin. The solution for the hyper-plane is found by solving the objective as follows:

$$\begin{aligned} & \underset{\mathbf{w}, \xi, \rho}{\text{minimize}} \quad \left( \frac{1}{2} \mathbf{w}^T \mathbf{w} - \frac{1}{\nu N} \sum_{i=1}^{N} \xi_i - \rho \right) \\ & \text{subject to } \mathbf{w}^T \phi(\mathbf{x}_i) \geq \rho - \xi_i, \;\; i = 1, \ldots, N \\ & \qquad \xi_i \geq 0, \end{aligned} \tag{2.39}$$

where $\mathbf{w}$ is the vector perpendicular to the separating hyper-plane (norm vector), $N$ is the number of training data, $\rho$ is the bias value parameterizes the hyper-plane, $\nu$ is a regularization variable penalizing the outliers in the training data, $\xi_i$ are the slack variables. Taking the conditions for optimality from the Lagrangian as

$$L(\mathbf{w},\xi,\rho;\alpha,\beta) = \frac{1}{2}\mathbf{w}^T\mathbf{w} + \frac{1}{\nu N}\sum_{i=1}^{N}\xi_i - \rho - \sum_{i=1}^{N}\alpha_i^T\left(w^T\phi(\mathbf{x}_i) - \rho + \xi_i\right) - \sum_{i=1}^{n}\beta_i^T\xi_i. \tag{2.40}$$

one obtains the dual problem as

$$\begin{aligned} &\underset{\alpha}{\text{minimize}} \;\; \alpha^T K \alpha \\ &\text{subject to} \;\; 0 \le \alpha_i \le \frac{1}{\nu N}, \\ &\qquad\qquad \sum_{i=1}^{N}\alpha_i = 1, \end{aligned} \tag{2.41}$$

where $K$ is the kernel matrix which elements are obtained by applying the kernel trick $\mathscr{K}(\mathbf{x},\mathbf{y}) = \phi(\mathbf{x})^T\phi(\mathbf{y})$, the dual problem of 1-SVM is again a problem similar to the Rayleigh quotient-type form. However, the optimal $\alpha$ cannot be solved as the eigenvalue problem because the constraint $\alpha^T\alpha = 1$ does not hold in (2.41). Instead, $\alpha$ is solved via convex optimization.

## 2.3 Summary

In this chapter we made a survey of several popular machine learning algorithms. The main finding was that these problems can all be simplified as the Generalized Rayleigh quotient form, given by

$$\underset{W}{\text{minimize}} \;\; (W^T P W)^{-1}(W^T Q W) \;\; \text{or} \;\; \underset{W}{\text{maximize}}(W^T P W)^{-1}(W^T Q W). \tag{2.42}$$

In Table 2.1 we summarize the objectives, the mappings and the constraints corresponding of these algorithms in terms of Generalized Rayleigh quotient.

**Table 2.1** Summary of several machine algorithms and the mappings to the Generalized Rayleigh quotient form

| **algorithm** | **objective** | **mapping** | **constraints** |
|---|---|---|---|
| PCA | $\max_{\mathbf{w}} \mathbf{w}^T C_{\mathbf{xx}} \mathbf{w}$ | $Q = C_{\mathbf{xx}},\ P = I,\ W = \mathbf{w}$ | $\mathbf{w}^T \mathbf{w} = 1$ |
| CCA | $\max_{\mathbf{w}_1,\mathbf{w}_2} \frac{\begin{bmatrix}\mathbf{w}_1\\ \mathbf{w}_2\end{bmatrix}^T \begin{bmatrix}0 & C_{\mathbf{xy}}\\ C_{\mathbf{yx}} & 0\end{bmatrix} \begin{bmatrix}\mathbf{w}_1\\ \mathbf{w}_2\end{bmatrix}}{\begin{bmatrix}\mathbf{w}_1\\ \mathbf{w}_2\end{bmatrix}^T \begin{bmatrix}C_{\mathbf{xx}} & 0\\ 0 & C_{\mathbf{yy}}\end{bmatrix} \begin{bmatrix}\mathbf{w}_1\\ \mathbf{w}_2\end{bmatrix}}$ | $Q = \begin{bmatrix}0 & C_{\mathbf{xy}}\\ C_{\mathbf{yx}} & 0\end{bmatrix}, P = \begin{bmatrix}C_{\mathbf{xx}} & 0\\ 0 & C_{\mathbf{yy}}\end{bmatrix}, W = \begin{bmatrix}\mathbf{w}_1\\ \mathbf{w}_2\end{bmatrix}$ | $\begin{bmatrix}\mathbf{w}_1\\ \mathbf{w}_2\end{bmatrix}^T \begin{bmatrix}\mathbf{w}_1\\ \mathbf{w}_2\end{bmatrix} = I$ |
| FDA | $\max_{\mathbf{w}} \left(\frac{\mathbf{w}^T S_B \mathbf{w}}{\mathbf{w}^T S_W \mathbf{w}}\right)$ | $Q = S_B,\ P = S_W,\ W = \mathbf{w}$ | $\mathbf{w}^T \mathbf{w} = I$ |
| $k$-means | $\max_A \operatorname{trace}\left(A^T X^T X A\right)$ | $Q = X^T X,\ P = I,\ W = A$ | $A^T A = I$ |
| Spectral Clustering | $\min_A \operatorname{trace}(A^T D^{-\frac{1}{2}} L D^{-\frac{1}{2}} A)$ | $Q = L,\ P = D,\ W = A$ | $A^T A = I$ |
| KL | $\max_A \operatorname{trace}\left(\frac{A^T X^T X A}{A^T L A}\right)$ | $Q = X^T X,\ P = L,\ A = P$ | $A^T A = I$ |
| 1-SVM | $\min_\alpha \alpha^T K \alpha$ | $Q = K,\ P = I,\ A = \alpha$ | $\sum_{i=1}^N \alpha_i = 1$ |

## References

1. Bhatia, R.: Matrix Analysis. Springer, New York (1997)
2. Chan, P.K., Schlag, M.D.F., Zien, J.Y.: Spectral K-way ratio-cut partitioning and clustering. In: Proc. of the 30th International Design Automation Conference, pp. 749–754. ACM Press, New York (1993)
3. Ding, C., He, X.: K-means Clustering via Principal Component Analysis. In: Proc. of ICML 2004, pp. 225–232. ACM Press, New York (2004)
4. Horn, R.A., Johnson, C.R.: Matrix Analysis. Cambridge University Press, Cambridge (1986)
5. Hotelling, H.: Relations between two sets of variates. Biometrika 28, 321–377 (1936)
6. Kovač-Striko, J., Veselić, K.: Trace Minimization and Definiteness of Symmetric Pencils. Linear algebra and its applications 216, 139–158 (1995)
7. von Luxburg, U.: A tutorial on spectral clustering. Statistics and Computing 17, 395–416 (2007)
8. Maila, M., Shi, J.: A Random Walks View of Spectral Segmentation. In: Proc. of AI and STATISTICS, AISTATS 2001(2001)
9. Mika, S., Weston, J., Schölkopf, B., Smola, A., Müller, K.-R.: Constructing Descriptive and Discriminative Nonlinear Features: Rayleigh Coefficients in Kernel Feature Spaces. IEEE Trans. PAMI 25, 623–628 (2003)
10. Nakić, I., Veselić, K.: Wielandt and Ky-Fan theorem for matrix pairs. Linear Algebra and its Applications 369, 177–193 (2003)
11. Shi, J., Malik, J.: Normalized cuts and image segmentation. IEEE Trans. PAMI 22, 888–905 (2000)
12. Stewart, G.W., Sun, J.-G.: Matrix Perturbation Theory. Academic Press, London (1990)
13. Wagner, D., Wagner, F.: Between min cut and graph bisection. In: Borzyszkowski, A.M., Sokolowski, S. (eds.) MFCS 1993. LNCS, vol. 711, pp. 744–750. Springer, Heidelberg (1993)
14. Wang, F., Ding, C., Li, T.: Integrated KL(K-means-Laplacian) Clustering: A New Clustering Approach by Combining Attribute Data and Pairwise Relations. In: SDM 2009, pp. 38–49. SIAM Press, Philadelphia (2009)
15. Zha, H., Ding, C., Gu, M., He, X., Simon, H.: Spectral relaxation for K-means clustering. Advances in Neural Information Processing Systems 13, 1057–1064 (2001)

# Chapter 3
# $L_n$-norm Multiple Kernel Learning and Least Squares Support Vector Machines

## 3.1 Background

In the era of information overflow, data mining and machine learning are indispensable tools to retrieve information and knowledge from data. The idea of incorporating several data sources in analysis may be beneficial by reducing the noise, as well as by improving statistical significance and leveraging the interactions and correlations between data sources to obtain more refined and higher-level information [50], which is known as *data fusion*. In bioinformatics, considerable effort has been devoted to *genomic data fusion*, which is an emerging topic pertaining to a lot of applications. At present, terabytes of data are generated by high-throughput techniques at an increasing rate. In data fusion, these terabytes are further multiplied by the number of data sources or the number of species. A statistical model describing this data is therefore not an easy matter. To tackle this challenge, it is rather effective to consider the data as being generated by a complex and unknown black box with the goal of finding a function or an algorithm that operates on an input to predict the output. About 15 years ago, Boser [8] and Vapnik [51] introduced the support vector method which makes use of kernel functions. This method has offered plenty of opportunities to solve complicated problems but also brought lots of interdisciplinary challenges in statistics, optimization theory, and the applications therein [40].

*Multiple kernel learning* (MKL) has been pioneered by Lanckriet *et al.* [29] and Bach *et al.* [6] as an additive extension of single kernel SVM to incorporate multiple kernels in classification. It has also been applied as a statistical learning framework for genomic data fusion [30] and many other applications [15]. The essence of MKL, which is the additive extension of the dual problem, relies only on the kernel representation (kernel trick) [3] while the heterogeneities of data sources are resolved by transforming different data structures (*e.g.*, vectors, strings, trees, graphs) into kernel matrices. In the dual problem, these kernels are combined into a single kernel, moreover, the coefficients of the kernels are leveraged adaptively to optimize the algorithmic objective, known as *kernel fusion*. The notion of kernel fusion was originally proposed to solve classification problems in computational biology, but recent efforts have lead to analogous solutions for one class [15] and unsupervised

S. Yu et al.: Kernel-based Data Fusion for Machine Learning, SCI 345, pp. 39–88.
springerlink.com 

learning problems [58]. Currently, most of the existing MKL methods are based on the formulation proposed by Lanckriet *et al.* [29], which is clarified in our paper as the optimization of the infinity norm ($L_\infty$) of kernel fusion. Optimizing $L_\infty$ MKL in the dual problem corresponds to posing $L_1$ regularization on the kernel coefficients in the primal problem. As known, $L_1$ regularization is characterized by the sparseness of the kernel coefficients [35]. Thus, the solution obtained by $L_\infty$ MKL is also sparse, which assigns dominant coefficients to only one or two kernels. The sparseness is useful to distinguish relevant sources from a large number of irrelevant data sources. However, in biomedical applications, there are usually a small number of sources and most of these data sources are carefully selected and preprocessed. They thus often are directly relevant to the problem. In these cases, a sparse solution may be too selective to thoroughly combine the complementary information in the data sources. While the performance on benchmark data may be good, the selected sources may not be as strong on truly novel problems where the quality of the information is much lower. We may thus expect the performance of such solutions to degrade significantly on actual real-world applications. To address this problem, we propose a new kernel fusion scheme by optimizing the $L_2$-norm of multiple kernels. The $L_2$ MKL yields a non-sparse solution, which smoothly distributes the coefficients on multiple kernels and, at the same time, leverages the effects of kernels in the objective optimization. Empirical results show that the $L_2$-norm kernel fusion can lead to a better performance in biomedical data fusion.

## 3.2 Acronyms

The symbols and notations used in this Chapter are defined in Table 1 (in the order of appearance).

**Table 3.1** Symbols used in Chapter 3

| | | |
|---|---|---|
| $\alpha$ | $\mathbb{R}^N$ | the dual variable of SVM |
| $Q$ | $\mathbb{R}^{N\times N}$ | a semi-positive definite matrix |
| $\mathscr{C}$ | $\mathbb{R}^N$ | a convex set |
| $\Omega$ | $\mathbb{R}^{N\times N}$ | a combination of multiple semi-positive definite matrices |
| $j$ | $\mathbb{N}$ | the index of kernel matrices |
| $p$ | $\mathbb{N}$ | the number of kernel matrices |
| $\theta$ | $[0,1]$ | coefficients of kernel matrices |
| $t$ | $[0,+\infty)$ | dummy variable in optimization problem |
| $\mathbf{s}$ | $\mathbb{R}^p$ | $\mathbf{s}=\{\alpha^T Q_1\alpha,...,\alpha^T Q_p\alpha\}^T$ |
| $\mathbf{v}$ | $\mathbb{R}^p$ | $\mathbf{v}=\{\alpha^T K_1\alpha,...,\alpha^T K_p\alpha\}^T$ |
| $\mathbf{w}$ | $\mathbb{R}^D$ or $\mathbb{R}^\Phi$ | the norm vector of the separating hyperplane |
| $\phi(\cdot)$ | $\mathbb{R}^D\rightarrow\mathbb{R}^\Phi$ | the feature map |
| $i$ | $\mathbb{N}$ | the index of training samples |
| $\mathbf{x}_i$ | $\mathbb{R}^D$ | the vector of the $i$-th training sample |
| $\rho$ | $\mathbb{R}$ | bias term in 1-SVM |
| $\nu$ | $\mathbb{R}^+$ | regularization term of 1-SVM |
| $\xi_i$ | $\mathbb{R}$ | slack variable for the $i$-th training sample |
| $K$ | $\mathbb{R}^{N\times N}$ | kernel matrix |
| $\mathscr{K}(\mathbf{x}_i,\mathbf{x}_j)$ | $\mathbb{R}^\Phi\times\mathbb{R}^\Phi\rightarrow\mathbb{R}$ | kernel function, $\mathscr{K}(\mathbf{x}_i,\mathbf{x}_j)=\phi(\mathbf{x}_i)^T\phi(\mathbf{x}_j)$ |
| $\mathbf{z}$ | $\mathbb{R}^D$ | the vector of a test data sample |
| $y_i$ | -1 or +1 | the class label of the $i-$th training sample |
| $Y$ | $\mathbb{R}^{N\times N}$ | the diagonal matrix of class labels $Y=diag(y_1,...,y_N)$ |
| $C$ | $\mathbb{R}^+$ | the box constraint on dual variables of SVM |
| $b$ | $\mathbb{R}^+$ | the bias term in SVM and LSSVM |
| $\gamma$ | $\mathbb{R}^p$ | $\gamma=\{\alpha^T YK_1Y\alpha,...,\alpha^T YK_pY\alpha\}^T$ |
| $k$ | $\mathbb{N}$ | the number of classes |
| $\eta$ | $\mathbb{R}^p$ | $\eta=\{\sum_{q=1}^k(\alpha_q^T Y_qK_1Y_q\alpha_q),...,\sum_{q=1}^k(\alpha_q^T Y_qK_pY_q\alpha_q)\}^T$ |
| $\delta$ | $\mathbb{R}^p$ | variable vector in SIP problem |
| $u$ | $\mathbb{R}$ | dummy variable in SIP problem |
| $q$ | $\mathbb{N}$ | the index of class number in classification problem, $q=1,...,k$ |
| $A$ | $\mathbb{R}^{N\times N}$ | $A_j=\sum_{q=1}^k(\alpha_q^T Y_qK_jY_q\alpha_q)$ |
| $\lambda$ | $\mathbb{R}^+$ | the regularization parameter in LSSVM |
| $e_i$ | $\mathbb{R}$ | the error term of the $i$-th sample in LSSVM |
| $\beta$ | $\mathbb{R}^N$ | the dual variable of LSSVM, $\beta=Y\alpha$ |
| $\varepsilon$ | $\mathbb{R}^+$ | precision value as the stopping criterion of SIP iteration |
| $\tau$ | $\mathbb{N}$ | index parameter of SIP iterations |
| $\mathbf{g}$ | $\mathbb{R}^p$ | $\mathbf{g}=\{\beta^T K_1\beta,...,\beta^T K_p\beta\}^T$ |
| $C_+,C_-$ | $\mathbb{R}^+$ | box constraints for weighted SVM |
| $\omega$ | $\mathbb{R}^N$ | vector of weights for weighted LSSVM |
| $W$ | $\mathbb{R}^{N\times N}$ | diagonal matrix of weights for weighted LSSVM |

## 3.3 The Norms of Multiple Kernel Learning

### 3.3.1 $L_\infty$-norm MKL

We consider the problem of minimizing a quadratic cost of a real vector in function of $\alpha$ and a real positive semi-definite (PSD) matrix $Q$, given by

$$\begin{aligned} &\underset{\alpha}{\text{minimize}} \;\; \alpha^T Q \alpha \\ &\text{subject to} \;\; \alpha \in \mathscr{C}, \end{aligned} \tag{3.1}$$

where $\mathscr{C}$ denotes a convex set. Also, PSD implies that $\forall \alpha$, $\alpha^T Q \alpha \geq 0$. We will show that many machine learning problems can be cast as the form in (3.1) with additional constraints on $\alpha$. In particular, if we restrict $\alpha^T \alpha = 1$, the problem in (3.1) becomes a Rayleigh quotient and leads to a eigenvalue problem.

Now we consider a convex parametric linear combination of a set of $p$ PSD matrices $Q_j$, given by

$$\Omega = \left\{ \sum_{j=1}^{p} \theta_j Q_j \middle| \forall j,\ \theta_j \geq 0,\ Q_j \succeq 0 \right\}. \tag{3.2}$$

To bound the coefficients $\theta_j$, we restrict that, for example, $||\theta_j||_1 = 1$, thus (3.1) can be equivalently rewritten as a min-max problem, given by

$$\begin{aligned} &\underset{\alpha}{\text{minimize}} \; \underset{\theta}{\text{maximize}} \;\; \alpha^T \left( \sum_{j=1}^{p} \theta_j Q_j \right) \alpha \\ &\text{subject to} \;\; Q_j \succeq 0, \;\; j = 1, ..., p \\ &\qquad\qquad\;\; \alpha \in \mathscr{C}, \\ &\qquad\qquad\;\; \theta_j \geq 0, \;\; j = 1, ..., p \\ &\qquad\qquad\;\; \sum_{j=1}^{p} \theta_j = 1. \end{aligned} \tag{3.3}$$

To solve (3.3), we denote $t = \alpha^T \left( \sum_{j=1}^{p} \theta_j Q_j \right) \alpha$, then the min-max problem can be formulated in a form of *quadratically constrained linear programming* (QCLP) [10], given by

$$\begin{aligned} &\underset{\alpha, t}{\text{minimize}} \;\; t \\ &\text{subject to} \;\; Q_j \succeq 0, \;\; j = 1, ..., p \\ &\qquad\qquad\;\; \alpha \in \mathscr{C}, \\ &\qquad\qquad\;\; t \geq \alpha^T Q_j \alpha, \;\; j = 1, ..., p\,. \end{aligned} \tag{3.4}$$

The optimal solution $\theta^*$ in (3.3) is obtained from the dual variable corresponding to the quadratic constraints in (3.4). The optimal $t^*$ is equivalent to the *Chebyshev* or $L_\infty$-norm of the vector of quadratic terms, given by

$$t^* = ||\alpha^T Q_j \alpha||_\infty = \max\{\alpha^T Q_1 \alpha, ..., \alpha^T Q_p \alpha\}. \tag{3.5}$$

The $L_\infty$-norm is the upper bound w.r.t. the constraint $\sum_{j=1}^p \theta_j = 1$ because

$$\alpha^T \left( \sum_{j=1}^p \theta_j Q_j \right) \alpha \leq t^*. \tag{3.6}$$

### 3.3.2 $L_2$-norm MKL

Apparently, suppose the optimal $\alpha^*$ is given, optimizing the $L_\infty$-norm in (3.5) will pick the single term with the maximal value, and the optimal solution of the coefficients is more likely to be sparse. An alternative solution to (3.3) is to introduce a different constraint on the coefficients, for example, $||\theta_j||_2 = 1$. We thus propose a new extension of the problem in (3.1), given by

$$\begin{aligned} \underset{\alpha}{\text{minimize}}\ \underset{\theta}{\text{maximize}}\ & \alpha^T \left( \sum_{j=1}^p \theta_j Q_j \right) \alpha \\ \text{subject to}\ & Q_j \succeq 0,\ \ j = 1, ..., p \\ & \alpha \in \mathscr{C}, \\ & \theta_j \geq 0,\ \ j = 1, ..., p \\ & ||\theta_j||_2 = 1. \end{aligned} \tag{3.7}$$

This new extension is analogously solved as a QCLP problem with modified constraints, given by

$$\begin{aligned} \underset{\alpha,\eta}{\text{minimize}}\ & \eta \\ \text{subject to}\ & Q_j \succeq 0,\ \ j = 1, ..., p \\ & \alpha \in \mathscr{C}, \\ & \eta \geq ||\mathbf{s}||_2,\ \ j = 1, ..., p, \end{aligned} \tag{3.8}$$

where $\mathbf{s} = \{\alpha^T Q_1 \alpha, ..., \alpha^T Q_p \alpha\}^T$. The proof that (3.8) is the solution of (3.7) is given in the following theorem.

**Theorem 3.1.** *The QCLP problem in (3.8) is the solution of the problem in (3.7).*

*Proof.* Given two vectors $\{x_1, ..., x_p\}, \{y_1, ..., y_p\}, x_j, y_j \in \mathbb{R}, j = 1, ..., p$, the Cauchy-Schwarz inequality states that

$$0 \leq \left( \sum_{j=1}^{p} x_j y_j \right)^2 \leq \sum_{j=1}^{p} x_j^2 \sum_{j=1}^{p} y_j^2, \tag{3.9}$$

with as equivalent form:

$$0 \leq \left[ \left( \sum_{j=1}^{p} x_j y_j \right)^2 \right]^{\frac{1}{2}} \leq \left[ \sum_{j=1}^{p} x_j^2 \sum_{j=1}^{p} y_j^2 \right]^{\frac{1}{2}}. \tag{3.10}$$

Let us denote $x_j = \theta_j$ and $y_j = \alpha^T Q_j \alpha$, (3.10) becomes

$$0 \leq \sum_{j=1}^{p} \left( \theta_j \alpha^T Q_j \alpha \right) \leq \left[ \sum_{j=1}^{p} \theta_j^2 \sum_{j=1}^{p} \left( \alpha^T Q_j \alpha \right)^2 \right]^{\frac{1}{2}}. \tag{3.11}$$

Since $||\theta_j||_2 = 1$, (3.11) is equivalent to

$$0 \leq \sum_{j=1}^{p} \left( \theta_j \alpha^T Q_j \alpha \right) \leq \left[ \sum_{j=1}^{p} \left( \alpha^T Q_j \alpha \right)^2 \right]^{\frac{1}{2}}. \tag{3.12}$$

Therefore, given $\mathbf{s} = \{\alpha^T Q_1 \alpha, ..., \alpha^T Q_p \alpha\}^T$, the additive term $\sum_{j=1}^{p} \left( \theta_j \alpha^T Q_j \alpha \right)$ is bounded by the $L_2$-norm $||\mathbf{s}||_2$. □

Moreover, it is easy to prove that when $\theta_j^* = \alpha^T Q_j \alpha / ||\mathbf{s}||_2$, the parametric combination reaches the upperbound and the equality holds. Optimizing this $L_2$-norm yields a non-sparse solution in $\theta_j$. In order to distinguish this from the solution obtained by (3.3) and (3.4), we denote it as the $L_2$-norm approach. It can also easily be seen (not shown here) that the $L_1$-norm approach is simply averaging the quadratic terms with uniform coefficients.

### 3.3.3 $L_n$-norm MKL

The $L_2$-norm bound is also generalizable to any positive real number $n \geq 1$, defined as $L_n$-norm MKL. Recently, the similar topic is also investigated by Kloft *et al.* [27] and a solution is proposed to solve the primal MKL problem. We will show that our primal-dual interpretation of MKL is also extendable to the $L_n$-norm. Let us assume that $\theta$ is regularized by the $L_m$-norm as $||\theta||_m = 1$, then the $L_m$-norm extension of equation (3.7) is given by

$$\begin{aligned} \underset{\alpha}{\text{minimize}}\ \underset{\theta}{\text{maximize}}\ & \alpha^T \left( \sum_{j=1}^{p} \theta_j Q_j \right) \alpha \qquad (3.13) \\ \text{subject to}\ & Q_j \succeq 0, \ \ j = 1, ..., p \\ & \alpha \in \mathscr{C}, \\ & \theta_j \geq 0, \ \ j = 1, ..., p \\ & ||\theta||_m = 1. \end{aligned}$$

In the following theorem, we prove that (3.13) can be equivalently solved as a QCLP problem, given by

$$\begin{aligned} \underset{\alpha,\eta}{\text{minimize}} \quad & \eta \\ \text{subject to} \quad & Q_j \succeq 0, \quad j=1,...,p \\ & \alpha \in \mathscr{C}, \\ & \eta \geq ||\mathbf{s}||_n, \end{aligned} \tag{3.14}$$

where $\mathbf{s} = \{\alpha^T Q_1 \alpha, ..., \alpha^T Q_p \alpha\}^T$ and the constraint is in $L_n$-norm, moreover, $n = \frac{m}{m-1}$. The problem in (3.14) is convex and can be solved by cvx toolbox [19, 20].

**Theorem 3.2.** *If the coefficient vector $\theta$ is regularized by a $L_m$-norm in (3.13), the problem can be solved as a convex programming problem in (3.14) with $L_n$-norm constraint. Moreover, $n = \frac{m}{m-1}$.*

*Proof.* We generalize the Cauchy-Schwarz inequality to Hölder's inequality. Let $m, n > 1$ be two numbers that satisfy $\frac{1}{m} + \frac{1}{n} = 1$. Then

$$0 \leq \sum_{j=1}^{p} x_j y_j \leq \left(\sum_{j=1}^{p} x_j^m\right)^{\frac{1}{m}} \left(\sum_{j=1}^{p} y_j^n\right)^{\frac{1}{n}}. \tag{3.15}$$

Let us denote $x_j = \theta_j$ and $y_j = \alpha^T Q_j \alpha$, (3.15) becomes

$$0 \leq \sum_{j=1}^{p} \left(\theta_j \alpha^T Q_j \alpha\right) \leq \left(\sum_{j=1}^{p} \theta_j^m\right)^{\frac{1}{m}} \left[\sum_{j=1}^{p} \left(\alpha^T Q_j \alpha\right)^n\right]^{\frac{1}{n}}. \tag{3.16}$$

Since $||\theta||_m = 1$, therefore the term $\left(\sum_{j=1}^{p} \theta_j^m\right)^{\frac{1}{m}}$ can be omitted in the equation, so (3.16) is equivalent to

$$0 \leq \sum_{j=1}^{p} \left(\theta_j \alpha^T Q_j \alpha\right) \leq \left[\sum_{j=1}^{p} \left(\alpha^T Q_j \alpha\right)^n\right]^{\frac{1}{n}}. \tag{3.17}$$

Due to the condition that $\frac{1}{m} + \frac{1}{n} = 1$, so $n = \frac{m}{m-1}$, we prove that with the $L_m$-norm constraint posed on $\theta$, the additive multiple kernel term $\sum_{j=1}^{p} \left(\theta_j \alpha^T Q_j \alpha\right)$ is bounded by the $L_n$-norm of the vector $\{\alpha^T Q_1 \alpha, ..., \alpha^T Q_n \alpha\}^T$. Moreover, we have $n = \frac{m}{m-1}$. □

In this section, we have explained the $L_\infty$, $L_1$, $L_2$, and $L_n$-norm approaches to extend the basic problem in (3.1) to multiple matrices $Q_j$. These approaches differed mainly on the constraints applied on the coefficients. To clarify the difference of notations used in this paper with the common interpretations of $L_1$ and $L_2$ regularization on $\theta$, we illustrate the mapping of our $L_\infty$, $L_1$, $L_2$, and $L_n$ notations between the common interpretations of coefficient regularization. As shown in Table 3.2, the notations used in this section are interpreted in the dual space and are equivalent to regularization of kernel coefficients in the primal space. The advantage of dual space interpretation is that we can easily extend the analogue solution to various machine learning algorithms, which have been shown in the previous chapter as the similar Rayleigh quotient problem.

**Table 3.2** This relationship between the norm of regularization constrained in the primal problem with the norm of kernels optimized in the dual problem

| | primal problem | dual problem |
|---|---|---|
| norm | $\theta_j$ | $\alpha^T K_j \alpha$ |
| $\mathbf{L}_\infty$ | $\lvert\theta\rvert = 1$ | $\max \lvert\lvert\{\alpha^T K_1 \alpha, ..., \alpha^T K_j \alpha\}\rvert\rvert_\infty$ |
| $\mathbf{L}_1$ | $\theta_j = \bar{\theta}$ | $\max \lvert\lvert\{\alpha^T K_1 \alpha, ..., \alpha^T K_j \alpha\}\rvert\rvert_1$ |
| $\mathbf{L}_2$ | $\lvert\lvert\theta\rvert\rvert_2 = 1$ | $\max \lvert\lvert\{\alpha^T K_1 \alpha, ..., \alpha^T K_j \alpha\}\rvert\rvert_2$ |
| $\mathbf{L}_{1.5}$ | $\lvert\lvert\theta\rvert\rvert_3 = 1$ | $\max \lvert\lvert\{\alpha^T K_1 \alpha, ..., \alpha^T K_j \alpha\}\rvert\rvert_{1.5}$ |
| $\mathbf{L}_{1.3333}$ | $\lvert\lvert\theta\rvert\rvert_4 = 1$ | $\max \lvert\lvert\{\alpha^T K_1 \alpha, ..., \alpha^T K_j \alpha\}\rvert\rvert_{1.3333}$ |
| $\mathbf{L}_{1.25}$ | $\lvert\lvert\theta\rvert\rvert_5 = 1$ | $\max \lvert\lvert\{\alpha^T K_1 \alpha, ..., \alpha^T K_j \alpha\}\rvert\rvert_{1.25}$ |
| $\mathbf{L}_{1.2}$ | $\lvert\lvert\theta\rvert\rvert_6 = 1$ | $\max \lvert\lvert\{\alpha^T K_1 \alpha, ..., \alpha^T K_j \alpha\}\rvert\rvert_{1.2}$ |
| $\mathbf{L}_{1.1667}$ | $\lvert\lvert\theta\rvert\rvert_7 = 1$ | $\max \lvert\lvert\{\alpha^T K_1 \alpha, ..., \alpha^T K_j \alpha\}\rvert\rvert_{1.1667}$ |

Next, we will investigate several concrete MKL algorithms and will propose the corresponding $L_2$-norm and $L_n$-norm solutions.

## 3.4 One Class SVM MKL

The primal problem of *one class SVM* (1-SVM) is defined by Tax and Duin [48] and Schölkopf *et al.* [38] as

$$\boxed{\text{P:}} \quad \begin{aligned} \underset{\mathbf{w},\xi,\rho}{\text{minimize}} \quad & \frac{1}{2}\mathbf{w}^T\mathbf{w} - \frac{1}{\nu N}\sum_{i=1}^{l}\xi_i - \rho \\ \text{subject to} \quad & \mathbf{w}^T\phi(\mathbf{x}_i) \geq \rho - \xi_i, \ i = 1, \ldots, N \\ & \xi_i \geq 0, \ i = 1, ..., N \end{aligned} \tag{3.18}$$

where $\mathbf{w}$ is the norm vector of the separating hyperplane, $\mathbf{x}_i$ are the training samples, $\nu$ is the regularization constant penalizing outliers in the training samples, $\phi(\cdot)$ denotes the feature map, $\rho$ is a bias term, $\xi_i$ are slack variables, and $N$ is the number of training samples. Taking the conditions for optimality from the Lagrangian, one obtains the dual problem, given by:

$$\boxed{\text{D:}} \quad \begin{aligned} \underset{\alpha}{\text{minimize}} \quad & \alpha^T K \alpha \\ \text{subject to} \quad & 0 \le \alpha_i \le \frac{1}{\nu N}, \quad i = 1, ..., N, \\ & \sum_{i=1}^{N} \alpha_i = 1, \end{aligned} \tag{3.19}$$

where $\alpha_i$ are the dual variables, $K$ represents the kernel matrix obtained by the inner product between any pair of samples specified by a kernel function $\mathcal{K}(\mathbf{x}_i, \mathbf{x}_j) = \phi(\mathbf{x}_i)^T \phi(\mathbf{x}_j),\ i, j = 1, ..., N$. To incorporate multiple kernels in (3.19), De Bie *et al.* proposed a solution [15] with the dual problem formulated as

$$\boxed{\text{D:}} \quad \begin{aligned} \underset{\alpha}{\text{minimize}} \quad & t \\ \text{subject to} \quad & t \ge \alpha^T K_j \alpha, \; j = 1, ..., p \\ & 0 \le \alpha_i \le \frac{1}{\nu N}, \; i = 1, ..., N \\ & \sum_{i=1}^{N} \alpha_i = 1, \end{aligned} \tag{3.20}$$

where $p$ is the number of data sources and $K_j$ is the $j$-th kernel matrix. The formulation exactly corresponds to the $L_\infty$ solution of the problem defined in the previous section (the PSD constraint is implied in the kernel matrix) with additional constraints imposed on $\alpha$. The optimal coefficients $\theta_j$ are used to combine multiple kernels as

$$\Omega = \left\{ \sum_{j=1}^{p} \theta_j K_j \,\middle|\, \sum_{j=1}^{p} \theta_j = 1, \;\; \forall j, \; \theta_j \ge 0 \right\}, \tag{3.21}$$

and the ranking function is given by

$$f(\mathbf{z}) = \frac{1}{\sqrt{\alpha^T \Omega_N \alpha}} \sum_{i=1}^{N} \alpha_i \Omega(\mathbf{z}, \mathbf{x}_i), \tag{3.22}$$

where $\Omega_N$ is the combined kernel of training data $\mathbf{x}_i,\ i = 1, ..., N$, $\mathbf{z}$ is the test data point to be ranked, $\Omega(\mathbf{z}, \mathbf{x}_i)$ is the combined kernel function applied on test data and training data, $\alpha$ is the dual variable solved as (3.20). De Bie *et al.* applied the method in the application of disease gene prioritization, where multiple genomic data sources are combined to rank a large set of test genes using the 1-SVM model trained from a small set of training genes which are known to be relevant for certain diseases. The $L_\infty$ formulation in their approach yields a sparse solution when integrating genomic data sources (see Figure two of [15]). To avoid this disadvantage, they proposed a regularization method by restricting the minimal boundary on the kernel coefficients, notated as $\theta_{min}$, to ensure the minimal contribution of each genomic data source to be $\theta_{min}/p$. According to their experiments, the regularized solution performed best, being significantly better than the sparse integration and the average combination of kernels.

Instead of setting the ad hoc parameter $\theta_{min}$, one can also straightforwardly propose an $L_2$-norm approach to solve the identical problem, given by

$$\boxed{\text{D:}} \quad \begin{aligned} &\underset{\alpha}{\text{minimize}} \ t \\ &\text{subject to} \ t \geq ||\mathbf{v}||_2, \\ &\qquad 0 \leq \alpha_i \leq \frac{1}{\nu N}, \ i = 1, ..., N \\ &\qquad \sum_{i=1}^{N} \alpha_i = 1, \end{aligned} \tag{3.23}$$

where $\mathbf{v} = \{\alpha^T K_1 \alpha, ..., \alpha^T K_p \alpha\}^T$, $\mathbf{v} \in \mathbb{R}^p$. The problem above is a QCLP problem and can be solved by conic optimization solvers such as Sedumi [39]. In (3.23), the first constraint represents a Lorentz cone and the second constraint corresponds to $p$ number of rotated Lorentz cones (R cones). The optimal kernel coefficients $\theta_j$ correspond to the dual variables of the R cones with $||\theta_j||_2 = 1$. In this $L_2$-norm approach, the integrated kernel $\Omega$ is combined by different $\theta_j^*$ and the same scoring function as in (3.22) is applied on the different solutions of $\alpha$ and $\Omega$.

## 3.5 Support Vector Machine MKL for Classification

### 3.5.1 The Conic Formulation

The notion of MKL is originally proposed in a binary SVM classification, where the primal objective is given by

$$\boxed{\text{P:}} \quad \begin{aligned} &\underset{\mathbf{w},b,\xi}{\text{minimize}} \ \frac{1}{2}\mathbf{w}^T\mathbf{w} + C\sum_{i=1}^{N} \xi_i \\ &\text{subject to} \ y_i[\mathbf{w}^T\phi(\mathbf{x}_i) + b] \geq 1 - \xi_i, \\ &\qquad i = 1, ..., N \\ &\qquad \xi_i \geq 0, \ i = 1, ..., N, \end{aligned} \tag{3.24}$$

where $\mathbf{x}_i$ are data samples, $\phi(\cdot)$ is the feature map, $y_i$ are class labels, $C > 0$ is a positive regularization parameter, $\xi_i$ are slack variables, $\mathbf{w}$ is the norm vector of the separating hyperplane, and $b$ is the bias. This problem is convex and can be solved as a dual problem, given by

$$\boxed{\text{D:}} \quad \begin{aligned} &\underset{\alpha}{\text{minimize}} \ \frac{1}{2}\alpha^T YKY\alpha - \alpha^T\mathbf{1} \\ &\text{subject to} \ (Y\alpha)^T\mathbf{1} = 0 \\ &\qquad 0 \leq \alpha_i \leq C, \ i = 1, ..., N, \end{aligned} \tag{3.25}$$

where $\alpha$ are the dual variables, $Y = diag(y_1, ..., y_N)$, $K$ is the kernel matrix, and $C$ is the upperbound of the box constraint on the dual variables. To incorporate multiple kernels in (3.25), Lanckriet *et al.* [29, 30] and Bach *et al.* [6] proposed a multiple kernel learning (MKL) problem as follows:

$$\boxed{\text{D:}} \quad \begin{aligned} \underset{t,\alpha}{\text{minimize}} \quad & \frac{1}{2}t - \alpha^T \mathbf{1} \\ \text{subject to} \quad & (Y\alpha)^T \mathbf{1} = 0 \\ & 0 \le \alpha_i \le C, \; i = 1, ..., N \\ & t \ge \alpha^T Y K_j Y \alpha, \;\; j = 1, ..., p, \end{aligned} \tag{3.26}$$

where $p$ is the number of kernels. The equation in (3.26) optimizes the $L_\infty$-norm of the set of kernel quadratic terms. Based on the previous discussions, its $L_2$-norm solution is analogously given by

$$\boxed{\text{D:}} \quad \begin{aligned} \underset{t,\alpha}{\text{minimize}} \quad & \frac{1}{2}t - \alpha^T \mathbf{1} \\ \text{subject to} \quad & (Y\alpha)^T \mathbf{1} = 0 \\ & 0 \le \alpha_i \le C, \;\; i = 1, ..., N \\ & t \ge ||\gamma||_2, \end{aligned} \tag{3.27}$$

where $\gamma = \{\alpha^T Y K_1 Y \alpha, ..., \alpha^T Y K_p Y \alpha\}^T$, $\gamma \in \mathbb{R}^p$. Both formulations in (3.26) and (3.27) can be efficiently solved as *second order cone programming* (SOCP) problems by a conic optimization solver (*e.g.*, Sedumi [39]) or as *Quadratically constrained quadratic programming* (QCQP) problems by a general QP solver (*e.g.*, MOSEK [4]). It is also known that a binary MKL problem can also be formulated as *Semi-definite Programming* (SDP), as proposed by Lanckriet *et al.* [29] and Kim *et al.* [25]. However, in a multi-class problem, SDP problems are computationally prohibitive due to the presence of PSD constraints and can only be solved approximately by relaxation [54]. On the contrary, the QCLP and QCQP formulations of binary classification problems can be easily extended to a multi-class setting using the one-versus-all (1vsA) coding, *i.e.*, solving the problem of $k$ classes as $k$ number of binary problems. Therefore, the $L_\infty$ multi-class SVM MKL is then formulated as

$$\boxed{\text{D:}} \quad \begin{aligned} \underset{t,\alpha}{\text{minimize}} \quad & \frac{1}{2}t - \sum_{q=1}^{k} \alpha_q^T \mathbf{1} \\ \text{subject to} \quad & (Y_q \alpha_q)^T \mathbf{1} = 0, \;\; q = 1, ..., k \\ & 0 \le \alpha_{iq} \le C, \;\; i = 1, ..., N, \\ & q = 1, ..., k \\ & t \ge \sum_{q=1}^{k} (\alpha_q^T Y_q K_j Y_q \alpha_q), \\ & j = 1, ..., p. \end{aligned} \tag{3.28}$$

The $L_2$ multi-class SVM MKL is given by

$$\boxed{\text{D:}} \quad \begin{aligned} \underset{t,\alpha}{\text{minimize}} \quad & \frac{1}{2}t - \sum_{q=1}^{k} \alpha_q^T \mathbf{1} \\ \text{subject to} \quad & (Y_q \alpha_q)^T \mathbf{1} = 0, \ \ q = 1,...,k, \\ & 0 \leq \alpha_{iq} \leq C, \ \ i = 1,...,N, \\ & q = 1,...,k \\ & t \geq ||\eta||_2, \end{aligned} \tag{3.29}$$

where
$\eta = \{\sum_{q=1}^{k}(\alpha_q^T Y_q K_1 Y_q \alpha_q),...,\sum_{q=1}^{k}(\alpha_q^T Y_q K_p Y_q \alpha_q)\}^T$, $\eta \in \mathbb{R}^p$.

### 3.5.2 The Semi Infinite Programming Formulation

Unfortunately, the kernel fusion problem becomes challenging on large scale data because it may scale up in three dimensions: the number of data points, the number of classes, and the number of kernels. When these dimensions are all large, memory issues may arise as the kernel matrices need to be stored in memory. Though it is feasible to approximate the kernel matrices by a low rank decomposition (*e.g.*, *incomplete Cholesky decomposition*) and to reduce the computational burden of conic optimization using these low rank matrices, conic problems involve a large amount of variables and constraints and it is usually less efficient than QCQP. Moreover, the precision of the low rank approximation relies on the assumption that the eigenvalues of kernel matrices decay rapidly, which may not always be true when the intrinsic dimensions of the kernels are large. To tackle the computational burden of MKL, Sonnenburg *et al.* reformulated the QP problem as *semi-infinite programming* (SIP) and approximated the QP solution using a bi-level strategy (wrapper method) [42]. The standard form of SIP is given by

$$\begin{aligned} \underset{\delta}{\text{maximize}} \quad & \mathbf{c}^T \delta \\ \text{subject to} \quad & f_t(\delta) \leq 0, \ \forall t \in \Upsilon, \end{aligned} \tag{3.30}$$

where the constraint functions in $f_t(\delta)$ can be either linear or quadratic and there are infinite number of them in $\forall t \in \Upsilon$. To solve it, a *discretization* method is usually applied, which is briefly summarized as follows [23, 22, 37]:

1. Choose a finite subset $\mathcal{N} \subset \Upsilon$.
2. Solve the convex programming problem

$$\underset{\delta}{\text{maximize}} \quad \mathbf{c}^T \delta \tag{3.31}$$

$$\text{subject to} \quad f_t(\delta) \leq 0, \ \ t \in \mathcal{N}. \tag{3.32}$$

3. If the solution of 2 is not satisfactorily close to the original problem then choose a larger, but still finite subset $\mathcal{N}$ and repeat from Step 2.

The convergence of SIP and the accuracy of the discretization method have been extensively described (*e.g.*, see [22, 23, 37]). As proposed by Sonnenburg *et al.* [42], the multi-class SVM MKL objective in (3.26) can be formulated as a SIP problem, given by

$$\begin{aligned}
\underset{\theta}{\text{maximize}} \quad & u \\
\text{subject to} \quad & \theta_j \geq 0, \; j = 1, ..., p \\
& \sum_{j=1}^{p} \theta_j = 1, \\
& \sum_{j=1}^{p} \theta_j f_j(\alpha_q) \geq u, \; \forall \alpha_q, \; q = 1, ..., k \\
& f_j(\alpha_q) = \sum_{q=1}^{k} \left( \frac{1}{2} \alpha_q^T Y_q K_j Y_q \alpha_q - \alpha_q^T \mathbf{1} \right), \\
& 0 \leq \alpha_{iq} \leq C, \; i = 1, ..., N, \; q = 1, ..., k \\
& (Y_q \alpha_q)^T \mathbf{1} = 0, \; q = 1, ..., k \, .
\end{aligned} \tag{3.33}$$

The SIP problem above is solved as a bi-level algorithm for which the pseudo code is presented in Algorithm 3.5.1.

---

**Algorithm 3.5.1.** SIP-SVM-MKL($K_j, Y_q, C, \varepsilon$)

---

Obtain the initial guess $\alpha^{(0)} = [\alpha_1, ..., \alpha_k]$
**while** $(\Delta u > \varepsilon)$

$$\textbf{do} \begin{cases} step1 : \text{Fix } \alpha, \text{ solve } \theta^{(\tau)} \text{then obtain } u^{(\tau)} \\ step2 : \text{Compute kernel combination } \Omega^{(\tau)} \\ step3 : \text{Solve single SVM by minimizing} f_j(\alpha_q) \text{ and obtain the optimal } \alpha_q^{(\tau)} \\ step4 : \text{Compute } f_1(\alpha^{(\tau)}), ..., f_p(\alpha^{(\tau)}) \\ step5 : \Delta u = |1 - \frac{\sum_{j=1}^{p} \theta_i^{(\tau-1)} f_j(\alpha^{(\tau)})}{u^{(\tau-1)}}| \end{cases}$$

**comment:** $\tau$ is the indicator of the current loop
**return** $(\theta^*, \alpha^*)$

---

In each loop $\tau$, Step 1 optimizes $\theta^{(\tau)}$ and $u^{(\tau)}$ for a restricted subset of constraints as a linear programming. Step 3 is an SVM problem with a single kernel and generates a new $\alpha^{(\tau)}$. If $\alpha^{(\tau)}$ is not satisfied by the current $\theta^{(\tau)}$ and $u^{(\tau)}$, it will be added successively to step 1 until all constraints are satisfied. The starting points $\alpha_q^{(0)}$ are randomly initialized and SIP always converges to a identical result.

Algorithm 3.5.1 is also applicable to the $L_2$-norm situation of SVM MKL, whereas the non-convex constraint $||\theta||_2 = 1$ in Step 1 needs to be relaxed as $||\theta||_2 \leq 1$, and the $f_j(\alpha)$ term in (3.32) is modified as only containing the quadratic term. The SIP formulation for $L_2$-norm SVM MKL is given by

$$\begin{aligned}
\underset{\theta,u}{\text{maximize}} \quad & u \\
\text{subject to} \quad & \theta_j \geq 0, \; j = 1,...,p, \\
& ||\theta||_2 \leq 1, \\
& \sum_{j=1}^{p} \theta_j f_j(\alpha_q) - \sum_{q=1}^{k} \alpha_q^T \mathbf{1} \geq u, \\
& \forall \alpha_q, \; q = 1,...,k \\
& f_j(\alpha_q) = \frac{1}{2} \sum_{q=1}^{k} \left(\alpha_q^T Y_q K_j Y_q \alpha_q\right), j = 1,...,p \\
& 0 \leq \alpha_{iq} \leq C, \; i = 1,...,N, \; q = 1,...,k \\
& (Y_q \alpha_q)^T \mathbf{1} = 0, \; q = 1,...,k.
\end{aligned} \tag{3.34}$$

With these modifications, Step 1 of Algorithm 3.5.1 becomes a QCLP problem given by

$$\begin{aligned}
\underset{\theta,u}{\text{maximize}} \quad & u \\
\text{subject to} \quad & \frac{1}{2} \sum_{j=1}^{p} \theta_j A_j - \alpha^T \mathbf{1} \geq u, \\
& 1 \geq \theta_1^2 + ... + \theta_p^2,
\end{aligned} \tag{3.35}$$

where $A_j = \sum_{q=1}^{k} \left(\alpha_q^T Y_q K_j Y_q \alpha_q\right)$ and $\alpha$ is a given value. Moreover, the PSD property of kernel matrices ensures that $A_j \geq 0$, thus the optimal solution always satisfies $||\theta||_2 = 1$. The extensions to the $L_n$-norm are also similar to this manner.

In the SIP formulation, the SVM MKL is solved iteratively as two components. The first component is a single kernel SVM, which is solved more efficiently when the data scale is larger then thousands of data points (and smaller than ten thousands) and, requires much less memory than the QP formulation. The second component is a small scale problem, which is a linear problem in $L_\infty$ case and a QCLP problem in the $L_2$ approach. As shown, the complexity of the SIP based SVM MKL is mainly determined by the burden of a single kernel SVM multiplied by the number of iterations. This has inspired us to adopt more efficient single SVM learning algorithms to further improve the efficiency. The least squares support vector machines (LSSVM) [45, 46, 47] is known for its simple differentiable cost function, the equality constraints in the separating hyperplane and its solution based on linear equations, which is preferable for large scaler problems. Next, we will investigate the MKL solutions issue using LSSVM formulations.

## 3.6 Least Squares Support Vector Machines MKL for Classification

### 3.6.1 The Conic Formulation

In LSSVM [45, 46, 47], the primal problem is

$$\boxed{\text{P:}} \quad \begin{aligned} &\underset{\mathbf{w},b,\mathbf{e}}{\text{minimize}} && \frac{1}{2}\mathbf{w}^T\mathbf{w} + \frac{1}{2}\lambda\mathbf{e}^T\mathbf{e} \\ &\text{subject to} && y_i[\mathbf{w}^T\phi(\mathbf{x}_i) + b] = 1 - e_i, \\ &&& i = 1,...,N, \end{aligned} \tag{3.36}$$

where most of the variables are defined in a similar way as in (3.24). The main difference is that the nonnegative slack variable $\xi$ is replaced by a squared error term $\mathbf{e}^T\mathbf{e}$ and the inequality constraints are modified as equality ones. Taking the conditions for optimality from the Lagrangian, eliminating $\mathbf{w}, \mathbf{e}$, defining $\mathbf{y} = [y_1,...,y_N]^T$ and $Y = diag(y_1,...,y_N)$, one obtains the following linear system [45]:

$$\boxed{\text{D:}} \quad \left[\begin{array}{c|c} 0 & \mathbf{y}^T \\ \hline \mathbf{y} & YKY + I/\lambda \end{array}\right] \left[\begin{array}{c} b \\ \hline \alpha \end{array}\right] = \left[\begin{array}{c} 0 \\ \hline \mathbf{1} \end{array}\right], \tag{3.37}$$

where $\alpha$ are unconstrained dual variables. Without the loss of generality, we denote $\beta = Y\alpha$ and rewrite (3.37) as

$$\boxed{\text{D:}} \quad \left[\begin{array}{c|c} 0 & \mathbf{1}^T \\ \hline \mathbf{1} & K + Y^{-2}/\lambda \end{array}\right] \left[\begin{array}{c} b \\ \hline \beta \end{array}\right] = \left[\begin{array}{c} 0 \\ \hline Y^{-1}\mathbf{1} \end{array}\right]. \tag{3.38}$$

In (3.38), we add an additional constraint as $Y^{-2} = I$ then the coefficient becomes a static value in the multi-class case. In 1vsA coding, (3.37) requires to solve $k$ number of linear problems whereas in (3.38), the coefficient matrix is only factorized once such that the solution of $\beta_q$ w.r.t. the multi-class label vectors $\mathbf{y}_q$ is very efficient to obtain. The constraint $Y^{-2} = I$ can be simply satisfied by assuming the class labels to be -1 and +1. Thus, from now on, we assume $Y^{-2} = I$ in the following discussion.

To incorporate multiple kernels in LSSVM classification, the $L_\infty$-norm approach is a QP problem, given by (assuming $Y^{-2} = I$)

$$\begin{aligned} &\underset{\alpha,t}{\text{minimize}} && \frac{1}{2}t + \frac{1}{2\lambda}\beta^T\beta - \beta^T Y^{-1}\mathbf{1} \\ &\text{subject to} && \sum_{i=1}^{N}\beta_i = 0, \\ &&& t \geq \beta^T K_j \beta, \ j = 1,...,p. \end{aligned} \tag{3.39}$$

The $L_2$-norm approach is analogously formulated as

$$\begin{aligned} \underset{\alpha,t}{\text{minimize}} \quad & \frac{1}{2}t + \frac{1}{2\lambda}\beta^T\beta - \beta^T Y^{-1}\mathbf{1} \\ \text{subject to} \quad & \sum_{i=1}^{N} \beta_i = 0, \\ & t \geq ||\mathbf{g}||_2, \ j = 1, ..., p, \end{aligned} \tag{3.40}$$

where $\mathbf{g} = \{\beta^T K_1 \beta, ..., \beta^T K_p \beta\}^T$, $\mathbf{g} \in \mathbb{R}^p$. The $\lambda$ parameter regularizes the squared error term in the primal objective in (3.36) and the quadratic term $\beta^T\beta$ in the dual problem. Usually, the optimal $\lambda$ needs to be selected empirically by cross-validation. In the kernel fusion of LSSVM, we can alternatively transform the effect of regularization as an identity kernel matrix in $\frac{1}{2}\beta^T\left(\sum_{j=1}^{p} K_j + \theta_{p+1} I\right)\beta$, where $\theta_{p+1} = 1/\lambda$. Then the MKL problem of combining $p$ kernels is equivalent to combining $p+1$ kernels where the last kernel is an identity matrix with the optimal coefficient corresponding to the $\lambda$ value. This method has been mentioned by Lanckriet *et al.* to tackle the estimation of the regularization parameter in the soft margin SVM [29]. It has also been used by Ye *et al.* to jointly estimate the optimal kernel for discriminant analysis [54]. Saving the effort of validating $\lambda$ may significantly reduce the model selection cost in complicated learning problems. By this transformation, the objective of LSSVM MKL becomes similar to that of SVM MKL with the main difference that the dual variables are unconstrained. Though (3.39) and (3.40) can in principle both be solved as QP problems by a conic solver or a QP solver, the efficiency of a linear solution of the LSSVM is lost. Fortunately, in an SIP formulation, the LSSVM MKL can be decomposed into iterations of the master problem of single kernel LSSVM learning, which is an unconstrained QP problem, and a coefficient optimization problem with very small scale.

### 3.6.2 The Semi Infinite Programming Formulation

The $L_\infty$-norm approach of multi-class LSSVM MKL is formulated as

$$\begin{aligned} \underset{\theta,u}{\text{maximize}} \quad & u \\ \text{subject to} \quad & \theta_j \geq 0, \ j = 1, ..., p+1 \\ & \sum_{j=1}^{p+1} \theta_j = 1, \\ & \sum_{j=1}^{p+1} \theta_j f_j(\beta_q) \geq u, \ \forall \beta_q, \ q = 1, ..., k \\ & f_j(\beta_q) = \sum_{q=1}^{k} \left(\frac{1}{2}\beta_q^T K_j \beta_q - \beta_q^T Y_q^{-1}\mathbf{1}\right), \\ & j = 1, ..., p+1, \ q = 1, ..., k. \end{aligned} \tag{3.41}$$

In the formulation above, $K_j$ represents the $j$-th kernel matrix in a set of $p+1$ kernels with the $(p+1)$-th kernel being the identity matrix. The $L_2$-norm LSSVM MKL is formulated as

$$\begin{aligned}
\underset{\theta,u}{\text{maximize}} \quad & u \qquad\qquad (3.42)\\
\text{subject to} \quad & \theta_j \geq 0, \; j=1,...,p+1\\
& \sum_{j=1}^{p+1} \theta_j^2 \leq 1,\\
& \sum_{j=1}^{p+1} \theta_j f_j(\beta_q) - \sum_{q=1}^{k} \beta_q^T Y_q^{-1} \mathbf{1} \geq u,\\
& \forall \beta_q, \; q=1,...,k\\
& f_j(\beta_q) = \sum_{q=1}^{k} \left( \frac{1}{2} \beta_q^T K_j \beta_q \right),\\
& j=1,...,p+1, \; q=1,...,k.
\end{aligned}$$

The pseudocode of $L_\infty$-norm and $L_2$-norm LSSVM MKL is presented in Algorithm 3.6.1.

---

**Algorithm 3.6.1.** SIP-LSSVM-MKL$(K_j, Y_q, \varepsilon)$

---

Obtain the initial guess $\beta^{(0)} = [\beta_1, ..., \beta_k]$
**while** $(\Delta u > \varepsilon)$
**do**
$$\begin{cases}
step1: \text{Fix } \beta, \text{ solve } \theta^{(\tau)} \text{then obtain } u^{(\tau)}\\
step2: \text{Compute kernel combination } \Omega^{(\tau)}\\
step3: \text{Solve single LSSVM and obtain the optimal } \beta^{(\tau)}\\
step4: \text{Compute } f_1(\beta^{(\tau)}), ..., f_{p+1}(\beta^{(\tau)})\\
step5: \Delta u = |1 - \frac{\sum_{j=1}^{p+1} \theta_i^{(\tau-1)} f_j(\beta^{(\tau)})}{u^{(\tau-1)}}|
\end{cases}$$
**comment:** $\tau$ is the indicator of the current loop
**return** $(\theta^*, \beta^*)$

---

In $L_\infty$ approach, Step 1 optimizes $\theta$ as a linear programming. In $L_2$ approach, Step 1 optimizes $\theta$ as a QCLP problem. Since the regularization coefficient is automatically estimated as $\theta_{p+1}$, Step 3 simplifies to a linear problem as

$$\left[\begin{array}{c|c} 0 & \mathbf{1}^T \\ \hline \mathbf{1} & \Omega^{(\tau)} \end{array}\right] \left[\begin{array}{c} b^{(\tau)} \\ \hline \beta^{(\tau)} \end{array}\right] = \left[\begin{array}{c} 0 \\ \hline Y^{-1}\mathbf{1} \end{array}\right], \qquad (3.43)$$

where $\Omega^{(\tau)} = \sum_{j=1}^{p+1} \theta_j^{(\tau)} K_j$.

## 3.7 Weighted SVM MKL and Weighted LSSVM MKL

### 3.7.1 *Weighted SVM*

The conventional SVM does not perform well in the presence of imbalanced data. Weighted SVM was proposed to cope with this problem [36, 52, 59]. As an extension from conventional SVM two different penalty constraints were introduced for the positive and negative classes. The optimization problem becomes,

$$\boxed{\text{P:}} \qquad \begin{aligned} \underset{\mathbf{w},b,\xi}{\text{minimize}} \quad & \frac{1}{2}\mathbf{w}^T\mathbf{w} + C_+ \sum_{\{i|y_i=+1\}} \xi_i^{k+} + C_- \sum_{\{i|y_i=-1\}} \xi_i^{k-} \\ \text{subject to} \quad & y_i[\mathbf{w}^T\phi(\mathbf{x}_i) + b] \geq 1 - \xi_i,\ i = 1,...,N \\ & \xi_i \geq 0,\ i = 1,...,N, \end{aligned} \tag{3.44}$$

where $\mathbf{x}_i$ are data samples, $\phi(\cdot)$ is the feature map, $y_i$ are class labels, $C_+$ and $C_-$ are respectively the penalty coefficients for positive class samples and negative class samples, $\xi_i$ are slack variables, $k+$ and $k-$ are respectively the numbers of slack variables for positive and negative class samples, $\mathbf{w}$ is the norm vector of the separating hyperplane, and $b$ is the bias. This problem is also convex and can be solved as a dual problem, given by

$$\boxed{\text{D:}} \qquad \begin{aligned} \underset{\alpha}{\text{minimize}} \quad & \frac{1}{2}\alpha^T Y K Y \alpha - \alpha^T \mathbf{1} \\ \text{subject to} \quad & (Y\alpha)^T\mathbf{1} = 0 \\ & 0 \leq \alpha_i \leq C_+,\ \{\forall i | y_i = +1\} \\ & 0 \leq \alpha_i \leq C_-,\ \{\forall i | y_i = -1\}, \end{aligned} \tag{3.45}$$

where $\alpha$ are the dual variables, $Y = diag(y_1,...,y_N)$, $K$ is the kernel matrix, and $C_+$ and $C_-$ are two different upperbounds of the box constraints on dual variables correspond to different classes. The value of $C_+$ and $C_-$ should be predetermined. In practical, one can fix $C_-$ and optimize the performance on training data by varying $C_+$ [59]. Suppose the dominant class is $+$, then its penalty value $C_+$ should be smaller than the value of rare class samples. In this chapter, the reported results on pregnancy data are obtained by $2C_+ = C_- = 2$.

### 3.7.2 *Weighted SVM MKL*

The MKL extension of Weighted SVM is analogous to the MKL extension of the unweighted SVM. The $L_\infty$ MKL for binary class SVM is given by

$$\boxed{\text{D:}} \quad \begin{aligned} \underset{\gamma,\alpha}{\text{minimize}} \quad & \frac{1}{2}\gamma - \alpha^T\mathbf{1} \\ \text{subject to} \quad & (Y\alpha)^T\mathbf{1} = 0 \\ & 0 \leq \alpha_i \leq C_+, \ \{\forall i | y_i = +1\} \\ & 0 \leq \alpha_i \leq C_-, \ \{\forall i | y_i = -1\} \\ & \gamma \geq \alpha^T Y K_j Y \alpha, \ \ j = 1, ..., p, \end{aligned} \tag{3.46}$$

where $p$ is the number of kernels.

The $L_2$-norm MKL is analogously given by

$$\boxed{\text{D:}} \quad \begin{aligned} \underset{\eta,\alpha}{\text{minimize}} \quad & \frac{1}{2}\eta - \alpha^T\mathbf{1} \\ \text{subject to} \quad & (Y\alpha)^T\mathbf{1} = 0 \\ & 0 \leq \alpha_i \leq C_+, \ \{\forall i | y_i = +1\} \\ & 0 \leq \alpha_i \leq C_-, \ \{\forall i | y_i = -1\} \\ & \eta \geq ||\gamma_j||_2, \ \ j = 1, ..., p \\ & \gamma_j \geq \alpha^T Y K_j Y \alpha. \ \ j = 1, ..., p. \end{aligned} \tag{3.47}$$

### 3.7.3 *Weighted LSSVM*

In LSSVM, the cost function can be extended to cope with imbalanced data, given by

$$\begin{aligned} \underset{\mathbf{w},b,\mathbf{e}}{\text{minimize}} \quad & \frac{1}{2}\mathbf{w}^T\mathbf{w} + \frac{1}{2}\lambda\sum_{i=1}^{N} v_i e_i^2 \\ \text{subject to} \quad & y_i[\mathbf{w}^T\phi(\mathbf{x}_i) + b] = 1 - e_i, \ i = 1, ..., N, \end{aligned} \tag{3.48}$$

where the main difference with unweighted LSSVM is that the least squares terms are weighted for different samples. Suppose $\omega = \{\omega_1, \omega_2, ..., \omega_N\}$ is a vector of weights associated with each sample, taking the conditions for optimality from the Lagrangian, eliminating $\mathbf{w}, \mathbf{e}$, defining $\mathbf{y} = [y_1, ..., y_N]^T$, $Y = diag(y_1, ..., y_N)$ and $W = diag(\omega_1^{-1}, ..., \omega_N^{-1})$, the weighted LSSVM can be solved as the following linear system [44]:

$$\left[\begin{array}{c|c} 0 & \mathbf{y}^T \\ \hline \mathbf{y} & YKY + W/\lambda \end{array}\right] \left[\begin{array}{c} b \\ \hline \alpha \end{array}\right] = \left[\begin{array}{c} 0 \\ \hline \mathbf{1} \end{array}\right]. \tag{3.49}$$

To improve the robustness of LSSVM when coping with the imbalanced data, a simple way to choose the weighting factors is [11]

$$\omega_i = \begin{cases} N/2N_+ & \text{if } y_i = +1 \\ N/2N_- & \text{if } y_i = -1 \end{cases}, \tag{3.50}$$

where $N_+$ and $N_-$ represent the number of positive and negative samples respectively. In this chapter, the reported results on pregnancy data are obtained by weighted LSSVM using the setting described above.

### 3.7.4 Weighted LSSVM MKL

To incorporate multiple kernels in Weighted LSSVM, the QCQP based $L_\infty$ solution is given by (assuming $Y^{-2} = I$)

$$\begin{aligned}
\underset{\alpha,t}{\text{minimize}} \quad & \frac{1}{2}t + \frac{1}{2\lambda}\beta^T W \beta - \beta^T Y^{-1}\mathbf{1} \\
\text{subject to} \quad & \sum_{i=1}^{N} \beta_i = 0, \\
& t \geq \beta^T K_j \beta, \; j = 1, ..., p.
\end{aligned} \tag{3.51}$$

where most of the variables are defined the same as in the unweighted version. The weight factor matrix $W$ is defined as same in (3.49). The $L_2$-norm approach is analogously formulated as

$$\begin{aligned}
\underset{\alpha,\eta}{\text{minimize}} \quad & \frac{1}{2}\eta + \frac{1}{2\lambda}\beta^T W \beta - \beta^T Y^{-1}\mathbf{1} \\
\text{subject to} \quad & \sum_{i=1}^{N} \beta_i = 0, \\
& s_j \geq \beta^T K_j \beta, \; j = 1, ..., p, \\
& \eta \geq ||s_j||_2, \; j = 1, ..., p.
\end{aligned} \tag{3.52}$$

The SIP based formulations for Weighted LSSVM MKL are analogous to the unweighted version, with the only difference that the single kernel weighted LSSVM is solved as the linear system defined in (3.49).

## 3.8 Summary of Algorithms

As discussed, the dual $L_2$ MKL solution can be extended to many machine learning problems. In principle, all MKL algorithms can be formulated in $L_\infty$, $L_1$, $L_2$, and $L_n$ forms and lead to different solutions. To validate the proposed approach, we implemented and compared 20 algorithms on various data sets. The summary of all implemented algorithms is presented in Table 3.3. These algorithms combine $L_\infty$, $L_1$, and $L_2$ MKL with 1-SVM, SVM, LSSVM, Weighted SVM and Weighted LSSVM. Though we mainly focus on $L_\infty$, $L_1$, and $L_2$ MKL methods, we also implement the $L_n$-norm MKL for 1-SVM, SVM, LS-SVM and Weighted SVM. These algorithms are applied on the four biomedical experimental data sets and the performance is systematically compared.

**Table 3.3** Summary of the implemented MKL algorithms

| Algorithm Nr. | Formulation Nr. | Name | References | Formulation | Equations |
|---|---|---|---|---|---|
| 1 | 1-A | 1-SVM $L_\infty$ MKL | [15] | SOCP | (3.20) |
| 1 | 1-B | 1-SVM $L_\infty$ MKL | [15] | QCQP | (3.20) |
| 2 | 2-A | 1-SVM $L_\infty$ (0.5) MKL | [15] | SOCP | (3.20) |
| 2 | 2-B | 1-SVM $L_\infty$ (0.5) MKL | [15] | QCQP | (3.20) |
| 3 | 3-A | 1-SVM $L_1$ MKL | [38, 48] | SOCP | (3.19) |
| 3 | 3-B | 1-SVM $L_1$ MKL | [38, 48] | QCQP | (3.19) |
| 4 | 4-A | 1-SVM $L_2$ MKL | novel | SOCP | (3.23) |
| 5 | 5-B | SVM $L_\infty$ MKL | [6, 29, 30] | QCQP | (3.26) |
| 5 | 5-C | SVM $L_\infty$ MKL | [42] | SIP | (3.33) |
| 6 | 6-B | SVM $L_\infty$ (0.5) MKL | novel | QCQP | (3.26) |
| 7 | 7-A | SVM $L_1$ MKL | [51] | SOCP | (3.25) |
| 7 | 7-B | SVM $L_1$ MKL | [29] | QCQP | (3.25) |
| 8 | 8-A | SVM $L_2$ MKL | novel | SOCP | (3.27) |
| 8 | 8-C | SVM $L_2$ MKL | [26] | SIP | (3.34) |
| 9 | 9-B | Weighted SVM $L_\infty$ MKL | novel | QCQP | (3.46) |
| 10 | 10-B | Weighted SVM $L_\infty$ (0.5) MKL | novel | QCQP | (3.46) |
| 11 | 11-B | Weighted SVM $L_1$ MKL | [36, 52, 59] | QCQP | (3.45) |
| 12 | 12-A | Weighted SVM $L_2$ MKL | novel | SOCP | (3.47) |
| 13 | 13-B | LSSVM $L_\infty$ MKL | [54] | QCQP | (3.39) |
| 13 | 13-C | LSSVM $L_\infty$ MKL | [54] | SIP | (3.41) |
| 14 | 14-B | LSSVM $L_\infty$ (0.5) MKL | novel | QCQP | (3.39) |
| 15 | 15-D | LSSVM $L_1$ MKL | [45] | linear | (3.38) |
| 16 | 16-B | LSSVM $L_2$ MKL | novel | SOCP | (3.40) |
| 16 | 16-C | LSSVM $L_2$ MKL | novel | SIP | (3.42) |
| 17 | 17-B | Weighted LSSVM $L_\infty$ MKL | novel | QCQP | (3.51) |
| 18 | 18-B | Weighted LSSVM $L_\infty$ (0.5) MKL | novel | QCQP | (3.51) |
| 19 | 19-D | Weighted LSSVM $L_1$ MKL | [44] | linear | (3.49) |
| 20 | 20-A | Weighted LSSVM $L_2$ MKL | novel | SOCP | (3.52) |

## 3.9 Numerical Experiments

### *3.9.1 Overview of the Convexity and Complexity*

We concluded the convexity and the time complexity of all proposed methods in Table 3.4. All problems proposed in this chapter are convex or can be transformed to a convex formulation by relaxation. The LSSVM SIP formulation has the lowest time complexity thus it is more preferable for large scale problems. We verified the efficiency in numerical experiment, which adopts two UCI digit recognition data sets (pen-digit and optical digit) to compare the computational time of the proposed algorithms.

### *3.9.2 QP Formulation Is More Efficient than SOCP*

We investigated the efficiency of various formulations to solve the 1-SVM MKL. As mentioned, the problems presented in (15) can be solved either as QCLP or as SOCP. We applied Sedumi [39] to solve it as SOCP and MOSEK to solve it as QCLP and SOCP. We found that solving the QP by MOSEK was most efficient

**Table 3.4** Convexity and complexity of MKL methods. $n$ is the number of samples, $p$ is the number of kernels, $k$ is the number of classes, $\tau$ is the number of iterations in SIP. The complexity of LSSVM SIP depends on the algorithms used to solve the linear system. For the conjugate gradient method, the complexity is between $O(n^{1.5})$ and $O(n^2)$ [45].

| Method | convexity | complexity |
|---|---|---|
| 1-SVM SOCP $L_\infty$,$L_2$ | convex | $O((p+n)^2n^{2.5})$ |
| 1-SVM QCQP $L_\infty$ | convex | $O(pn^3)$ |
| (weighted) SVM SOCP $L_\infty$,$L_2$ | convex | $O((p+n)^2(k+n)^{2.5})$ |
| (weighted) SVM QCQP $L_\infty$ | convex | $O(pk^2n^2+k^3n^3)$ |
| (weighted) SVM SIP $L_\infty$ | convex | $O(\tau(kn^3+p^3))$ |
| (weighted) SVM SIP $L_2$ | relaxation | $O(\tau(kn^3+p^3))$ |
| (weighted) LSSVM SOCP $L_\infty$,$L_2$ | convex | $O((p+n)^2(k+n)^{2.5})$ |
| (weighted) LSSVM QCQP $L_\infty$,$L_2$ | convex | $O(pk^2n^2+k^3n^3)$ |
| (weighted) LSSVM SIP $L_\infty$ | convex | $O(\tau(n^2+p^3))$ |
| (weighted) LSSVM SIP $L_2$ | relaxation | $O(\tau(n^2+p^3))$ |

(142 seconds). In contrast, the MOSEK-SOCP method costed 2608 seconds and the Sedumi-SOCP method took 4500 seconds, shown in Figure 3.1. This is probably because when transforming a QP to a SOCP, a large number of additional variables and constraints are involved, thus becoming more expensive to solve.

### 3.9.3 SIP Formulation Is More Efficient than QCQP

To compare the computational time of solving MKL classifiers based on QP and SIP formulations, we scaled up the kernel fusion problem in three dimensions: the number of kernels, the number of classes and the number of samples. As shown in Figure 3.2 and 3.3, the SIP formulation of LSSVM MKL increases linearly with the number of samples and kernels, and is barely influenced by the number of classes (Figure 3.4). Solving the SIP based LSSVM MKL is significantly faster than solving SVM MKL because the former optimizes through iterations on a linear systems whereas the latter iterates over quadratic systems. For LSSVM MKL, the SIP formulation is also more preferable than the quadratic formulation. A quadratic system is a memory intensive problem and its complexity increases exponentially with the number of kernels and the number of samples in MKL. In contrast, the SIP formulation separates the problem into a series of linear systems, whose complexity is only determined by the number of samples and less affected by the number of kernels or classes. As shown in step 3 of Algorithm 3.6.1, the coefficient matrix of the linear system is a combined single kernel matrix and is constant with respect to multiple classes, thus it can be solved very efficiently. We have also compared the CPU time of $L_\infty$ and $L_2$ LSSVM MKL on large data sets and their efficiency is very similar to each other.

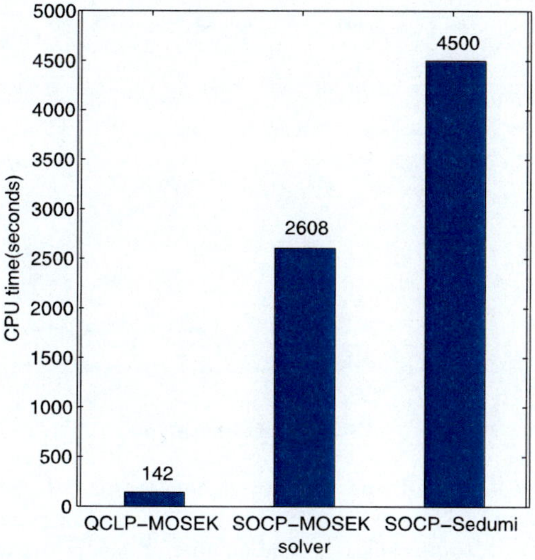


**Fig. 3.1** Comparison of SOCP and QCQP formulations to solve 1-SVM MKL using two kernels. To simulate the ranking problem in 1-SVM, 3000 digit samples were retrieved as training data. Two kernels were constructed respectively for each data source using RBF kernel functions. The computational time was thus evaluated by combining the two 3000 × 3000 kernel matrices.

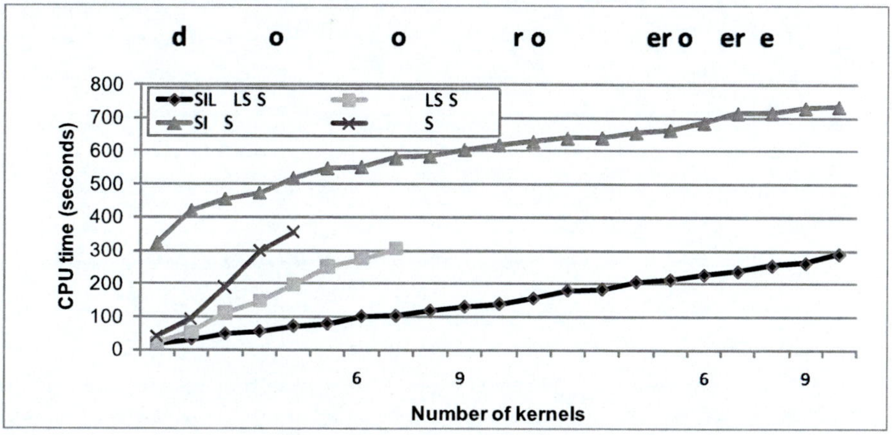


**Fig. 3.2** Comparison of the QCQP and the SIP formulations to solve the SVM MKL and the LSSVM MKL using different numbers of kernels. The benchmark data set was constructed by 2000 samples labeled in 2 classes. We used different kernel widths to construct the RBF kernel matrices and increase the number of kernel matrices from 2 to 200. The QCQP formulations had memory issues when the number of kernels was larger than 60. The experiment was carried on a dual Opteron 250 Unix system with 16Gb memory.

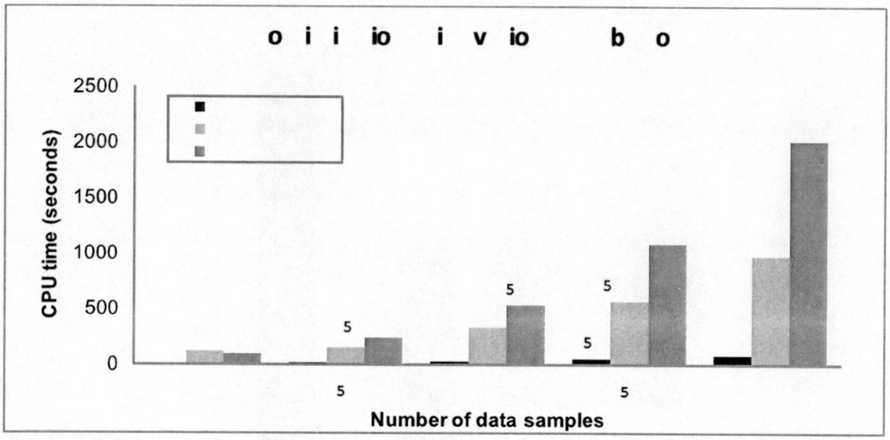


**Fig. 3.3** Comparison of the QCQP and SIP formulations to solve the SVM MKL and the LSSVM MKL using different sizes of samples. The benchmark data set was made up of two linear kernels and labels in 10 digit classes. The number of data points was increased from 1000 to 3000. The experiment was carried on a dual Opteron 250 Unix system with 16Gb memory.

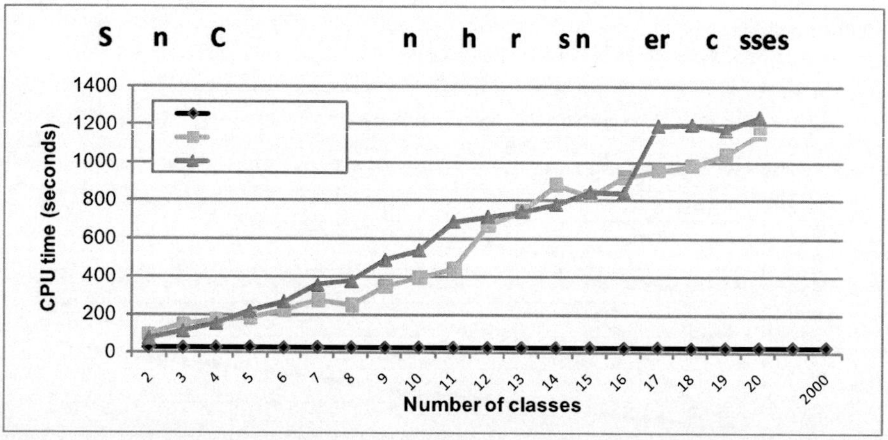


**Fig. 3.4** Comparison of the QCQP and SIP formulations to solve the SVM MKL and the LSSVM MKL data sets containing different numbers of classes. The benchmark data was made up of two linear kernel matrices and 2000 samples. The samples were equally and randomly divided into various number of classes. The class number gradually increased from 2 to 20. The experiment was carried on a dual Opteron 250 Unix system with 16Gb memory.

## 3.10 MKL Applied to Real Applications

The performance of the proposed $L_2$-norm MKL and $L_n$-norm method was systematically evaluated and compared on six real benchmark data sets. On each data set, we compared the $L_2$ and $L_n$ method with the $L_\infty$, $L_1$ and regularized $L_\infty$ MKL method. In the regularized $L_\infty$, we set the minimal boundary of kernel coefficients $\theta_{min}$ to 0.5, denoted as $L_\infty(0.5)$. The experiments were categorized in four groups as summarized in Table 3.5.

**Table 3.5** Summary of data sets and algorithms used in five experiments

| Nr. | Data Set | Problem | Samples | Classes | Algorihtms | Evaluation |
|---|---|---|---|---|---|---|
| 1 | disease relevant genes | ranking | 620 | 1 | 1-4 | LOO AUC |
| 2 | prostate cancer genes | ranking | 9 | 1 | 1-4 | AUC |
| 3 | rectal cancer patients | classification | 36 | 2 | 5-8,13-16 | LOO AUC |
| 4 | endometrial disease | classification | 339 | 2 | 5-8,13-16 | 3-fold AUC |
| | miscarriage | classification | 2356 | 2 | 5-8,13-16 | 3-fold AUC |
| | pregnancy | classification | 856 | 2 | 9-12,17-20 | 3-fold AUC |

### *3.10.1 Experimental Setup and Data Sets*

#### Experiment 1: Disease Relevant Gene Prioritization by Genomic Data Fusion

In the first experiment, we demonstrated a disease gene prioritization application to compare the performance of optimizing different norms in MKL. The computational definition of gene prioritization is mentioned in our earlier work [1, 15, 57]. We applied four 1-SVM MKL algorithms to combine kernels derived from 9 heterogeneous genomic sources (shown in section 1 of Additional file 1) to prioritize 620 genes that are annotated to be relevant for 29 diseases in OMIM. The performance was evaluated by leave-one-out (LOO) validation: for each disease which contains $K$ relevant genes, one gene, termed the "defector" gene, was removed from the set of training genes and added to 99 randomly selected test genes (test set). We used the remaining $K-1$ genes (training set) to build our prioritization model. Then, we prioritized the test set of 100 genes with the trained model and determined the rank of that defector gene in test data. The prioritization function in (22) scored the relevant genes higher and others lower, thus, by labeling the "defector" gene as class "+1" and the random candidate genes as class "-1", we plotted the Receiver Operating Characteristic (ROC) curves to compare different models using the error of AUC (one minus the area under the ROC curve).

The kernels of data sources were all constructed using linear functions except the sequence data that was transformed into a kernel using a 2-mer string kernel function [31] (details shown in Table 3.6).

In total 9 kernels were combined in this experiment. The regularization parameter $\nu$ in 1-SVM was set to 0.5 for all comparing algorithms. Since there was no

**Table 3.6** Genomic data sources used in experiment 1 and 2

| data source | reference | type | features | kernel function |
|---|---|---|---|---|
| EST | [17] | expressed sequence tagging annotations | 167 | linear |
| GO | [5] | GO annotations | 8643 | linear |
| Interpro | [34] | annotations | 4708 | linear |
| KEGG pathways | [24] | interactions | 314 | linear |
| Motif | [2, 32] | motif findings | 674 | linear |
| Sequence | [53] | amino acid sequences | 20 | 2-mer string |
| Microarray Son *et al.* | [41] | expression array | 158 | linear |
| Microarray Su *et al.* | [43] | expression array | 158 | linear |
| Text | [55, 57] | gene by term vectors using GO vocabulary | 7403 | linear |

hyper-parameter needed to be tuned in LOO validation, we reported the LOO results as the performance of generalization. For each disease relevant gene, the 99 test genes were randomly selected in each LOO validation run from the whole human protein-coding genome. We repeated the experiment 20 times and the mean value and standard deviation were used for comparison.

### Experiment 2: Prioritization of Recently Discovered Prostate Cancer Genes by Genomic Data Fusion

In the second experiment we used the same data sources and kernel matrices as in the previous experiment to prioritize 9 prostate cancer genes recently discovered by Eeles *et al.* [16], Thomas *et al.* [49] and Gudmundsson *et al.* [21]. A training set of 14 known prostate cancer genes was compiled from the reference database OMIM including only the discoveries prior to January 2008. This training set was then used to train the prioritization model. For each novel prostate cancer gene, the test set contained the newly discovered gene plus its 99 closest neighbors on the chromosome. Besides the error of AUC, we also compared the ranking position of the novel prostate cancer gene among its 99 closet neighboring genes. Moreover, we compared the MKL results with the ones obtained via the Endeavour application[1].

### Experiment 3: Clinical Decision Support by Integrating Microarray and Proteomics Data

The third experiment is taken from the work of Daemen *et al.* about the kernel-based integration of genome-wide data for clinical decision support in cancer diagnosis [14]. Thirty-six patients with rectal cancer were treated by combination of cetuximab, capecitabine and external beam radiotherapy and their tissue and plasma samples were gathered at three time points: before treatment ($T_0$); at the early therapy treatment ($T_1$) and at the moment of surgery ($T_2$). The tissue samples were hybridized to gene chip arrays and after processing, the expression was reduced to 6,913 genes. Ninety-six proteins known to be involved in cancer were measured in the plasma samples, and the ones that had absolute values above the detection limit in less than 20% of the samples were excluded for each time point separately. This resulted in the exclusion of six proteins at $T_0$ and four at $T_1$. "responders" were distinguished from "non-responders" according to the pathologic lymph node stage at surgery (pN-STAGE). The "responder" class contains 22 patients with no lymph node found at surgery whereas the "non-responder" class contains 14 patients with at least 1 regional lymph node. Only the two array-expression data sets (MA) measured at $T_0$ and $T_1$ and the two proteomics data sets (PT) measured at $T_0$ and $T_1$ were used to predict the outcome of cancer at surgery.

Similar to the original method applied on the data [14], we used R BioConductor DEDS as feature selection techniques for microarray data and the Wilcoxon rank sum test for proteomics data. The statistical feature selection procedure was independent to the classification procedure, however, the performance varied widely with the number of selected genes and proteins. We considered the relevance of features (genes and proteins) as prior knowledge and systematically evaluated the performance using multiple numbers of genes and proteins. According to the ranking of statistical feature selection, we gradually increased the number of genes and proteins from 11 to 36, and combined the linear kernels constructed by these features. The performance was evaluated by LOO method, where the reason was two folded: Firstly, the number of samples was small (36 patients); secondly, the kernels were all constructed with a linear function. Moreover, in LSSVM classification we proposed the strategy to estimate the regularization parameter $\lambda$ in kernel fusion. Therefore, no hyperparameter was needed to be tuned so we reported the LOO validation result as the performance of generalization.

#### Experiment 4: Clinical Decision Support by Integrating Multiple Kernels

Our fourth experiment considered three clinical data sets. These three data sets were derived from different clinical studies and were used by Daemen and De Moor [13] as validation data for clinical kernel function development. Data set I contains clinical information on 402 patients with an endometrial disease who underwent an echographic examination and color Droppler [7]. The patients are divided into two groups according to their histology: malignant (hyperplasia, polyp, myoma, and carcinoma) versus benign (proliferative endometrium, secretory endometrium, atrophia). After excluding patients with incomplete data, the data contains 339 patients of which 163 malignant and 176 benign. Data set II comes from a prospective observational study of 1828 women undergoing transvaginal sonography before 12 weeks gestation, resulting in data for 2356 pregnancies of which 1458 normal at week 12 and 898 miscarriages during the first trimester [9]. Data set III contains data on 1003 pregnancies of unknown location (PUL) [18]. Within the PUL group, there are four clinical outcomes: a failing PUL, an intrauterine pregnancy (IUP), an ectopic pregnancy (EP) or a persisting PUL. Because persisting PULs are rare (18 cases in the data set), they were excluded, as well as pregnancies with missing data. The final data set consists of 856 PULs among which 460 failing PULs, 330 IUPs, and 66 EPs. As the most important diagnostic problem is the correct classification of the EPs versus non-EPs [12], the data was divided as 790 non-EPs and 66 EPs.

To simulate a problem of combining multiple sources, for each data we created eight kernels and combined them using MKL algorithms for classification. The eight kernels included one linear kernel, three RBF kernels, three polynomial kernels and a clinical kernel. The kernel width of the first RBF kernel is selected by empirical rules as four times the average covariance of all the samples, the second and the third kernel widths were respectively six and eight times the average covariance. The degrees of the three polynomial kernels were set to 2, 3, and 4 respectively. The bias term of polynomial kernels was set to 1. The clinical kernels were constructed as proposed by Daemen and De Moor [14]. Let $\mathcal{K}_v(i,j)$ denotes the kernel function for variable $v$ between patients $i$ and $j$, $\mathcal{K}(i,j)$ represents the global, heterogeneous kernel matrix:

- *Continuous and ordinal clinical variables:* The same kernel function is proposed for these variable types:

$$\mathcal{K}_v(i,j) = \frac{C - |v_i - v_j|}{C}, \tag{3.53}$$

  where the constant value $C$ is usually defined as the range between maximal value between minimal value of variable $v$ on the training set, given by

$$C = max - min. \tag{3.54}$$

- *Nominal clinical variables:* For nominal variables, the kernel function between patients $i$ and $j$ is defined as

$$\mathcal{K}_v(i,j) = \begin{cases} 1 \text{ if } v_i = v_j \\ 0 \text{ if } v_i \neq v_j \end{cases}. \tag{3.55}$$

- *Final kernel for clinical data:* Because each individual kernel matrix has been normalized to the interval [0,1], the global, heterogeneous kernel matrix can be defined as the sum of the individual kernel matrices, divided by the total number of clinical variables. This matrix then describes the similarity for a class of patients based on a set of variables of different type.

For example, in the endometrial data set, we would like to calculate the kernel function between two patients $i$ and $j$ for the variables age, number of miscarriages/abortions, and menopausal status, which are respectively continuous, ordinal and nominal variables. Suppose that patient $i$ is 23 years old, has 1 miscarriage and the nominal menopausal status value is 2; patient $j$ is 28 years old, has 2 miscarriage and the menopausal status value is 3. Suppose that, based on the training data, the minimal age is 20 and the maximal age is 100. The minimal miscarriage number is 0 and the maximal number is 5. Then for each variable, the kernel functions between $i$ and $j$ are:

$$\mathcal{K}_{age}(i,j) = ((100-20)-|23-28|)/(100-20) = 0.9375 ,$$
$$\mathcal{K}_{miscarriage}(i,j) = ((5-0)-|1-2|)/(5-0) = 0.8 ,$$
$$\mathcal{K}_{menopausal}(i,j) = 0 .$$

The overall kernel function between patient $i$ and $j$ is given by

$$\mathcal{K}(i,j) = \frac{1}{3}(\mathcal{K}_{age} + \mathcal{K}_{miscarriage} + \mathcal{K}_{menopausal}) = 0.5792 .$$

We noticed that the class labels of the pregnancy data were quite imbalanced (790 non-EPs and 66 EPs). In literature, the class imbalanced problem can be tackled by modifying the cost of different classes in the objective function of SVM. Therefore, we applied weighted SVM MKL and weighted LSSVM MKL on the imbalanced pregnancy data. For the other two data sets, we compared the performance of SVM MKL and LSSVM MKL with different norms.

The performance of classification was benchmarked using 3-fold cross validation. Each data set was randomly and equally divided into 3 parts. As introduced in previous sections, when combining multiple pre-constructed kernels in LSSVM based algorithms, the regularization parameter $\lambda$ can be jointly estimated as the coefficient of an identity matrix. In this case we don't need to optimize any hyper-parameter in the LSSVM. In the estimation approach of LSSVM and all approaches of SVM, we therefore could use both training and validation data to train the classifier, and test data to evaluate the performance. The evaluation was repeated three times, so each part was used once as test data. The average performance was reported as the evaluation of one repetition. In the standard validation approach of LSSVM, each dataset was partitioned randomly into three parts for training, validation and testing. The classifier was trained on the training data and the hyper-parameter $\lambda$ was tuned on the validation data. When tuning the $\lambda$, its values were sampled uniformly on the log scale from $2^{-10}$ to $2^{10}$. Then, at optimal $\lambda$, the classifier was retrained on the combined training and validation set and the resulting model is tested on the testing set. Obviously, the estimation approach is more efficient than the validation approach because the former approach only requires one training process whereas the latter needs to perform 22 times an additional training (21 $\lambda$ values plus the model retraining). The performance of these two approaches was also investigated in this experiment.

### 3.10.2 Results

#### Experiment 1: Disease Relevant Gene Prioritization by Genomic Data Fusion

In the first experiment, the $L_2$ 1-SVM MKL algorithm performed the best (Error 0.0780). As shown in Table 3.7, the $L_\infty$ and $L_1$ approaches all performed significantly worse than the $L_2$ approach. For example, in the current experiment, when setting the minimal boundary of the kernel coefficients to 0.5, each data source was ensured to have a minimal contribution in integration, thereby improving the $L_\infty$ performance from 0.0923 to 0.0806, although still lower than $L_2$.

**Table 3.7** Results of experiment 1: prioritization of 620 disease relevant genes by genomic data fusion. The error of AUC values is evaluated by LOO validation in 20 random repetitions. The best performance ($L_2$) is shown in bold. The p-values are compared with the best performance using a paired t-test. As shown, the $L_2$ method is significantly better than other methods.

| | Error of AUC (mean) | Error of AUC (std.) | p-value |
|---|---|---|---|
| $L_\infty$ | 0.0923 | 0.0035 | $2.98 \cdot 10^{-17}$ |
| $L_\infty(0.5)$ | 0.0806 | 0.0033 | $2.66 \cdot 10^{-06}$ |
| $L_1$ | 0.0908 | 0.0042 | $1.92 \cdot 10^{-16}$ |
| $L_2$ | **0.0780** | 0.0034 | - |
| $L_{1.5}$ | 0.0865 | 0.0046 | $3.54 \cdot 10^{-07}$ |
| $L_{1.3333}$ | 0.0889 | 0.0047 | $7.83 \cdot 10^{-09}$ |
| $L_{1.25}$ | 0.0903 | 0.0047 | $3.41 \cdot 10^{-12}$ |
| $L_{1.2}$ | 0.0912 | 0.0048 | $6.49 \cdot 10^{-14}$ |
| $L_{1.1667}$ | 0.0919 | 0.0048 | $2.63 \cdot 10^{-17}$ |

In Figure 3.5 we illustrate the optimal kernel coefficients of different approaches. As shown, the $L_\infty$ method assigned dominant coefficients to Text mining and Gene Ontology data, whereas other data sources were almost discarded from integration. In contrast, the $L_2$ approach evenly distributed the coefficients over all data sources and thoroughly combined them in integration. When combining multiple kernels, sparse coefficients combine the model only with one or two kernels, making the combined model fragile with respect to the uncertainty and novelty. In real problems, the relevance of a new gene to a certain disease may not have been investigated thus a model solely based on Text and GO annotation is less reliable. $L_2$ based integration evenly combines multiple genomic data sources. In this experiment, the $L_2$ approach showed the same effect as the regularized $L_\infty$ by setting some minimal boundaries on kernel coefficients. However, in the regularized $L_\infty$, the minimal boundary $\theta_{min}$ usually is predefined according to the "rule of thumb". The main advantage of the $L_2$ approach is that the $\theta_{min}$ values are determined automatically for different kernels and the performance is shown to be better with the manually selected values.

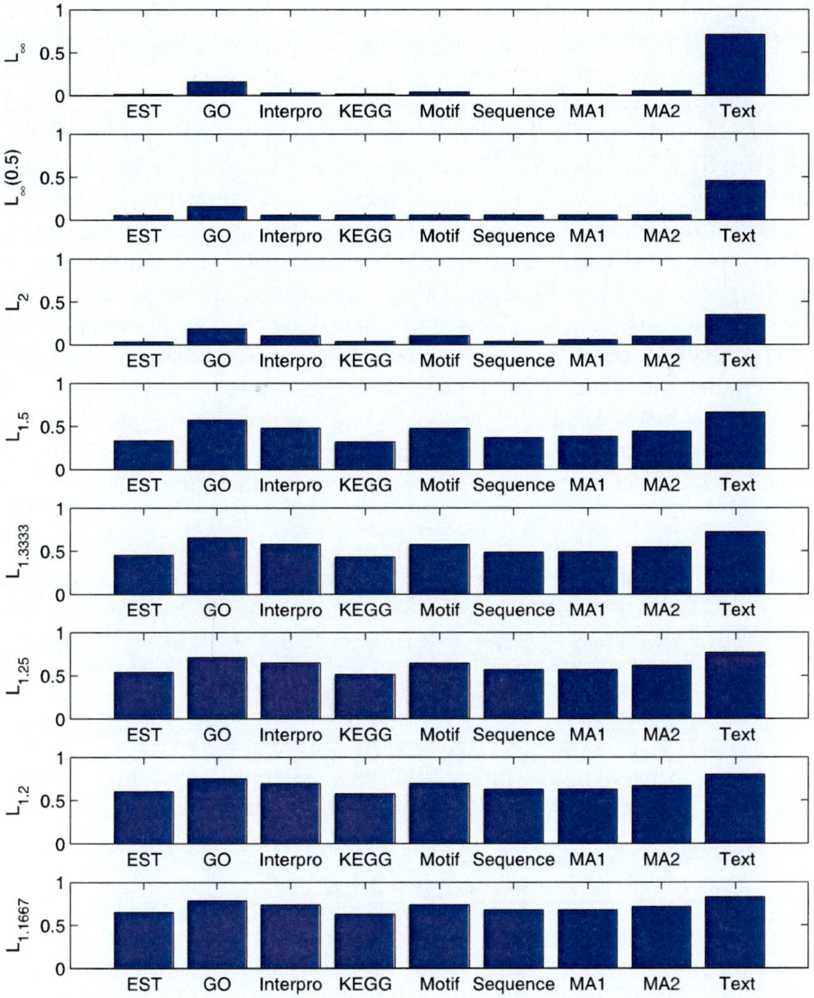


**Fig. 3.5** Optimal kernel coefficients assigned on genomic data sources in disease gene prioritization. For each method, the average coefficients of 20 repetitions are shown. The three most important data sources ranked by $L_\infty$ are Text, GO, and Motif. The coefficients on other six sources are almost zero. The $L_2$ method shows the same ranking on these three best data sources as $L_\infty$, moreover, it also shows ranking for other six sources. Thus, as another advantage of $L_2$ method, it provides more refined ranking of data sources than $L_\infty$ method in data integration. The coefficients optimized by some $L_n$-norm MKL are also illustrated. As shown, when $n$ approaches to 1, the coefficients become more evenly distributed on data sources.

### Experiment 2: Prioritization of Recently Discovered Prostate Cancer Genes by Genomic Data Fusion

In the second experiment, recently discovered prostate cancer genes were prioritized using the same data sources and algorithms as in the first experiment. As shown in Table 3.6, the $L_2$ method significantly outperformed other methods on prioritization of gene CDH23 [49], and JAZF1 [49]. For 5 other genes (CPNE [49], EHBP1 [21], MSMB [16], KLK3 [16], IL16 [49]), the performance of the $L_2$ method was comparable to the best result. In Table 3.8, we also presented the optimal kernel coefficients and the prioritization results for individual sources. As shown, the $L_\infty$ algorithm assigned most of the coefficients to Text and Microarray data. Text data performs well in the prioritization of known disease genes, however, does not always work the best for newly discovered genes. This experiment demonstrates that when prioritizing novel prostate cancer relevant genes, the $L_2$ MKL approach evenly optimized the kernel coefficients to combine heterogeneous genomic sources and its performance was significantly better than the $L_\infty$ method. Moreover, we also compared the kernel based data fusion approach with the Endeavour gene prioritization software: for 6 genes the MKL approach performed significantly better than Endeavour.

**Table 3.8** Results of experiment 2 with other norms: prioritization of prostate cancer genes by genomic data fusion. For each novel prostate cancer gene, the first row shows the error of AUC values and the second row lists the ranking position of the prostate cancer gene among its 99 closet neighboring genes.

| Name | $L_\infty$ | $L_\infty(0.5)$ | $L_1$ | $L_2$ | $L_{1.5}$ | $L_{1.3333}$ | $L_{1.25}$ | $L_{1.2}$ | $L_{1.1667}$ | Endeavour |
|---|---|---|---|---|---|---|---|---|---|---|
| CPNE | 0.3030 | 0.2323 | **0.1010** | *0.1212* | 0.1111 | 0.1111 | 0.1111 | 0.1111 | 0.1212 | - |
| | 31/100 | 24/100 | **11/100** | *13/100* | 12/10 | 12/10 | 12/10 | 12/10 | 13/10 | 70/100 |
| CDH23 | 0.0606 | 0.0303 | *0.0202* | **0.0101** | 0.0202 | 0.0202 | 0.0202 | 0.0202 | 0.0202 | - |
| | 7/100 | 4/100 | *3/100* | **2/100** | 3/10 | 3/10 | 3/10 | 3/10 | 3/10 | 78/100 |
| EHBP1 | 0.5354 | 0.5152 | **0.3434** | *0.3939* | 0.3737 | 0.3636 | 0.3535 | 0.3535 | 0.3535 | - |
| | 54/100 | 52/100 | **35/100** | *40/100* | 38/100 | 37/100 | 36/100 | 36/100 | 36/100 | 57/100 |
| MSMB | **0.0202** | **0.0202** | 0.0505 | *0.0303* | 0.0404 | 0.0505 | 0.0505 | 0.0505 | 0.0505 | - |
| | **3/100** | **3/100** | 6/100 | *4/100* | 5/100 | 6/100 | 6/100 | 6/100 | 6/100 | 69/100 |
| KLK3 | 0.3434 | 0.3535 | *0.2929* | *0.2929* | 0.3030 | 0.3030 | 0.3030 | 0.3030 | 0.3030 | - |
| | 35/100 | 36/100 | *30/100* | *30/100* | 31/100 | 31/100 | 31/100 | 31/100 | 31/100 | **28/100** |
| JAZF1 | *0.0505* | **0.0202** | **0.0202** | **0.0202** | **0.0202** | **0.0202** | **0.0202** | **0.0202** | **0.0202** | - |
| | *6/100* | **3/100** | **3/100** | **3/100** | **3/100** | **3/100** | **3/100** | **3/100** | **3/100** | 7/100 |
| LMTK2 | *0.3131* | 0.4646 | 0.8081 | 0.7677 | 0.7879 | 0.8081 | 0.8081 | 0.8081 | 0.8081 | - |
| | *32/100* | 47/100 | 81/100 | 77/100 | 78/100 | 79/100 | 81/100 | 81/100 | 81/100 | 81/100 |
| IL16 | **0** | *0.0101* | 0.0303 | *0.0101* | 0.0202 | 0.0303 | 0.0303 | 0.0303 | 0.0303 | - |
| | **1/100** | *2/100* | 4/100 | *2/100* | 3/100 | 4/100 | 4/100 | 4/100 | 4/100 | 72/100 |
| CTBP2 | 0.8283 | 0.5758 | *0.6364* | 0.6869 | 0.6667 | 0.6566 | 0.6465 | 0.6465 | 0.6465 | - |
| | 83/100 | 58/100 | *64/100* | 69/100 | 67/100 | 66/100 | 65/100 | 65/100 | 65/100 | **38/100** |

**Table 3.9** Results of experiment 2: prioritization of prostate cancer genes by genomic data fusion. For each gene, the best individual data sources are shown in bold. Apparently, the sparse kernel coefficients optimized by $L_\infty$ 1-SVM MKL is too selective thus sometimes the best individual data sources are discarded in integration. In comparison, the $L_2$ method is good at evenly combining multiple data sources.

| Name | | EST | GO | Interpro | KEGG | Motif | Sequence | MA1 (Son *et al.*) | MA2 (Su *et al.*) | Text |
|---|---|---|---|---|---|---|---|---|---|---|
| CPNE | Error AUC | 0.5000 | 0.2929 | 0.5000 | 0.4091 | 0.5000 | 0.3535 | 0.0909 | 0.3636 | **0.0505** |
| | Rank position | 58/100 | 30/100 | 37/100 | 50/100 | 53/100 | 36/100 | 10/100 | 37/100 | **6/100** |
| | $L_\infty$ coefficients | 0 | 0 | 0 | 0 | 0 | 0 | 0 | 0.7561 | 0.2439 |
| | $L_2$ coefficients | 0.0776 | 0.3006 | 0.2726 | 0.1423 | 0.2786 | 0.1789 | 0.4075 | 0.5400 | 0.4917 |
| CDH23 | Error AUC | 0.5000 | 0.1212 | 0.2929 | 0.5000 | 0.5000 | 0.2020 | 0.2929 | **0.0202** | 0.0606 |
| | Rank position | 71/100 | 13/100 | 30/100 | 89/100 | 54/100 | 21/100 | 30/100 | **3/100** | 7/100 |
| | $L_\infty$ coefficients | 0 | 0 | 0 | 0 | 0 | 0 | 0 | 0 | 1 |
| | $L_2$ coefficients | 0.0580 | 0.3331 | 0.3143 | 0.1972 | 0.2978 | 0.1823 | 0.2743 | 0.4412 | 0.5968 |
| EHBP1 | Error AUC | 0.5000 | 0.5000 | 0.2424 | 0.4545 | 0.4040 | 0.0707 | **0.0505** | 0.1414 | 0.5000 |
| | Rank position | 54/100 | 65/100 | 11/100 | 50/100 | 41/100 | 8/100 | **6/100** | 15/100 | 84/100 |
| | $L_\infty$ coefficients | 0 | 0 | 0 | 0 | 0 | 0 | 0 | 0.1905 | 0.8095 |
| | $L_2$ coefficients | 0.0733 | 0.3638 | 0.3013 | 0.1832 | 0.2921 | 0.1653 | 0.3833 | 0.4619 | 0.5148 |
| MSMB | Error AUC | 0.1616 | 0.3737 | 0.5000 | 0.5000 | 0.0606 | 0.3030 | **0.0202** | 0.3333 | 0.0303 |
| | Rank position | 15/100 | 38/100 | 60/100 | 92/100 | 7/100 | 31/100 | **3/100** | 34/100 | 4/100 |
| | $L_\infty$ coefficients | 0 | 0 | 0 | 0 | 0 | 0 | 1 | 0 | 0 |
| | $L_2$ coefficients | 0.0949 | 0.2936 | 0.2014 | 0.1198 | 0.2242 | 0.1256 | 0.7389 | 0.3198 | 0.3683 |
| KLK3 | Error AUC | 0.1616 | 0.5000 | **0.2475** | 0.4545 | 0.5000 | 0.5000 | 0.3535 | 0.3535 | 0.5000 |
| | Rank position | 17/100 | 63/100 | **19/100** | 20/100 | 87/100 | 77/100 | 36/100 | 36/100 | 94/100 |
| | $L_\infty$ coefficients | 0 | 0 | 0 | 0 | 0 | 0 | 0.9373 | 0 | 0.0627 |
| | $L_2$ coefficients | 0.1666 | 0.2534 | 0.3097 | 0.1193 | 0.2247 | 0.2345 | 0.5921 | 0.4149 | 0.4120 |
| JAZF1 | Error AUC | 0.4242 | 0.1212 | 0.5000 | 0.4444 | **0.0606** | 0.3131 | 0.2828 | 0.0707 | 0.1010 |
| | Rank position | 43/100 | 13/100 | 86/100 | 37/100 | **7/100** | 32/100 | 29/100 | 8/100 | 11/100 |
| | $L_\infty$ coefficients | 0 | 0 | 0 | 0 | 0 | 0 | 0.3541 | 0 | 0.6459 |
| | $L_2$ coefficients | 0.1301 | 0.2813 | 0.2990 | 0.1645 | 0.3004 | 0.1924 | 0.4661 | 0.4300 | 0.5081 |
| LMTK2 | Error AUC | 0.5000 | 0.5000 | 0.1212 | 0.4293 | 0.5000 | **0** | 0.5000 | 0.5000 | 0.3232 |
| | Rank position | 79/100 | 58/100 | 13/100 | 29/100 | 70/100 | **1/100** | 71/100 | 65/100 | 24/100 |
| | $L_\infty$ coefficients | 0 | 0 | 0 | 0 | 0 | 0 | 0 | 0 | 1 |
| | $L_2$ coefficients | 0.1134 | 0.3117 | 0.2836 | 0.1875 | 0.2777 | 0.2029 | 0.3584 | 0.4305 | 0.5850 |
| IL16 | Error AUC | **0** | 0.3939 | 0.5000 | 0.5000 | 0.5000 | 0.3636 | 0.0202 | 0.1212 | 0.0202 |
| | Rank position | **1/100** | 40/100 | 73/100 | 91/100 | 51/100 | 37/100 | 3/100 | 13/100 | 3/100 |
| | $L_\infty$ coefficients | 0 | 0 | 0 | 0 | 0 | 0 | 0.4991 | 0.2808 | 0.2201 |
| | $L_2$ coefficients | 0.0937 | 0.3723 | 0.2379 | 0.1456 | 0.2035 | 0.1502 | 0.5162 | 0.4918 | 0.4500 |
| CTBP2 | Error AUC | 0.5000 | **0.0808** | 0.5000 | 0.4546 | 0.5000 | 0.1313 | 0.5000 | 0.5000 | 0.5000 |
| | Rank position | 69/100 | **9/100** | 45/100 | 49/100 | 51/100 | 14/100 | 61/100 | 48/100 | 72/100 |
| | $L_\infty$ coefficients | 0 | 0 | 0 | 0 | 0 | 0 | 0 | 0.7801 | 0.2199 |
| | $L_2$ coefficients | 0.0726 | 0.3348 | 0.2368 | 0.1391 | 0.2419 | 0.1375 | 0.4189 | 0.5802 | 0.4665 |

### Experiment 3: Clinical Decision Support by Integrating Microarray and Proteomics Data

The $L_2$ MKL notion can be applied on various machine learning problems. The first two experiments demonstrated a ranking problem using 1-SVM MKL to prioritize disease relevant genes. In the third experiment we optimized the $L_\infty$, $L_1$, $L_2$, and $L_n$-norm in SVM MKL and LSSVM MKL classifiers to support the diagnosis of patients according to their lymph node stage in rectal cancer development. The performance of the classifiers greatly depended on the selected features, therefore, for each classifier we compared 25 feature selection results (as a grid of 5 numbers of genes multiplied by 5 numbers of proteins). As shown in Table 3.10, the best performance was obtained with LSSVM $L_1$ (error of AUC=0.0325) using 25 genes and 15 proteins. The $L_2$ LSSVM MKL classifier was also promising because its performance was comparable to the best result. In particular, for the two compared classifiers (LSSVM and SVM), the $L_1$ and $L_2$ approaches significantly outperformed the $L_\infty$ approach.

In LSSVM, the regularization parameter $\lambda$ was estimated jointly as the kernel coefficient of an identity matrix. In LSSVM $L_1$, $\lambda$ was set to 1. In all SVM approaches, the $C$ parameter of the box constraint was set to 1. In the table, the row and column labels represent the numbers of genes (g) and proteins (p) used to construct the kernels. The genes and proteins were ranked by feature selection techniques (see text). The AUC of LOO validation was evaluated without the bias term $b$ (as the implicit bias approach) because its value varied by each left out sample. In this problem, considering the bias term decreased the AUC performance. The performance was compared among eight algorithms for the same number of genes and proteins, where the best values (the smallest Error of AUC) are represented in bold, the second best ones in italic. The best performance of all the feature selection results is underlined. The table presents the 25 best feature selection results of each method.

We also tried other $L_n$-norms in the same experimental settings and the results are shown in Table 3.11. We found on some norms some specific results are further improved (*e.g.*, the combination of 15 proteins with 28 genes) but generally their performance is similar to the $L_2$ approach.

We also tried to regularize the kernel coefficients in $L_\infty$ MKL using different $\theta_{min}$ values. Nine different $\theta_{min}$ were tried uniformly from 0.1 to 0.9 and the changes in performance is shown in Figure 3.6. As shown, increasing the $\theta_{min}$ value steadily improves the performance of LSSVM MKL and SVM MKL on the rectal cancer data sets. However, determining the optimal $\theta_{min}$ was a non-trivial issue. When $\theta_{min}$ was smaller than 0.6, the performance of LSSVM MKL $L_\infty$ remained unchanged, meaning that the "rule of thumb" value 0.5 used in experiment 1 is not valid here. In comparison, when using the $L_2$ based MKL classifiers, there is no need to specify $\theta_{min}$ and the performance is still comparable to the best performance obtained with regularized $L_\infty$ MKL.

**Table 3.10** Results of experiment 3 using $L_\infty$, $L_1$, and $L_2$ MKL: classification of patients in rectal cancer clinical decision using microarray and proteomics data sets

| | LSSVM $L_\infty$ | | | | | SVM $L_\infty$ | | | | |
|---|---|---|---|---|---|---|---|---|---|---|
| | 14p | 15p | 16p | 17p | 18p | 14p | 15p | 16p | 17p | 18p |
| 24g | 0.0584 | 0.0519 | *0.0747* | 0.0812 | 0.0812 | 0.1331 | 0.1331 | 0.1331 | 0.1331 | 0.1364 |
| 25g | *0.0390* | *0.0390* | 0.0519 | 0.0617 | 0.0649 | 0.1136 | 0.1104 | 0.1234 | 0.1201 | 0.1234 |
| 26g | 0.0487 | 0.0487 | 0.0812 | 0.0844 | 0.0877 | 0.1266 | 0.1136 | 0.1234 | 0.1299 | 0.1364 |
| 27g | 0.0617 | 0.0649 | 0.0812 | 0.0877 | 0.0942 | 0.1429 | 0.1364 | 0.1364 | 0.1331 | 0.1461 |
| 28g | 0.0552 | 0.0487 | 0.0617 | 0.0747 | 0.0714 | 0.1429 | 0.1331 | 0.1331 | 0.1364 | 0.1396 |
| | LSSVM $L_\infty(0.5)$ | | | | | SVM $L_\infty(0.5)$ | | | | |
| | 14p | 15p | 16p | 17p | 18p | 14p | 15p | 16p | 17p | 18p |
| 24g | 0.0584 | 0.0519 | *0.0747* | 0.0812 | 0.0812 | 0.1266 | 0.1006 | 0.1266 | 0.1299 | 0.1331 |
| 25g | *0.0390* | *0.0390* | 0.0519 | 0.0617 | 0.0649 | 0.1136 | 0.1071 | 0.1234 | 0.1201 | 0.1234 |
| 26g | 0.0487 | 0.0487 | 0.0812 | 0.0844 | 0.0877 | 0.1136 | 0.1136 | 0.1201 | 0.1266 | 0.1331 |
| 27g | 0.0617 | 0.0649 | 0.0812 | 0.0877 | 0.0942 | 0.1364 | 0.1364 | 0.1364 | 0.1331 | 0.1461 |
| 28g | 0.0552 | 0.0487 | 0.0617 | 0.0747 | 0.0714 | 0.1299 | 0.1299 | 0.1299 | 0.1331 | 0.1364 |
| | LSSVM $L_1$ | | | | | SVM $L_1$ | | | | |
| | 14p | 15p | 16p | 17p | 18p | 14p | 15p | 16p | 17p | 18p |
| 24g | **0.0487** | **0.0487** | **0.0682** | **0.0682** | 0.0747 | 0.0747 | 0.0584 | 0.0714 | **0.0682** | 0.0747 |
| 25g | **0.0357** | **0.0325** | **0.0422** | **0.0455** | **0.0455** | 0.0584 | 0.0519 | 0.0649 | 0.0714 | 0.0714 |
| 26g | **0.0357** | **0.0357** | **0.0455** | **0.0455** | **0.0455** | 0.0584 | 0.0519 | 0.0682 | 0.0682 | 0.0682 |
| 27g | **0.0357** | **0.0357** | **0.0455** | **0.0487** | **0.0519** | 0.0617 | 0.0584 | 0.0714 | 0.0682 | 0.0682 |
| 28g | **0.0422** | **0.0325** | **0.0487** | **0.0487** | **0.0519** | 0.0584 | 0.0584 | 0.0649 | 0.0649 | 0.0682 |
| | LSSVM $L_2$ | | | | | SVM $L_2$ | | | | |
| | 14p | 15p | 16p | 17p | 18p | 14p | 15p | 16p | 17p | 18p |
| 24g | *0.0552* | **0.0487** | *0.0747* | *0.0779* | **0.0714** | 0.0909 | 0.0877 | 0.0974 | 0.0942 | 0.1006 |
| 25g | *0.0390* | *0.0390* | *0.0487* | *0.0552* | *0.0552* | 0.0747 | 0.0649 | 0.0812 | 0.0844 | 0.0844 |
| 26g | *0.0390* | *0.0455* | *0.0552* | *0.0649* | *0.0649* | 0.0747 | 0.0584 | 0.0812 | 0.0779 | 0.0779 |
| 27g | *0.0422* | *0.0487* | *0.0552* | *0.0584* | *0.0649* | 0.0779 | 0.0812 | 0.0844 | 0.0812 | 0.0812 |
| 28g | *0.0455* | **0.0325** | **0.0487** | *0.0584* | *0.0552* | 0.0812 | 0.0714 | 0.0812 | 0.0779 | 0.0812 |

**Table 3.11** Results of experiment 3 with other norms: classification of patients in rectal cancer clinical decision using microarray and proteomics data sets

| | LSSVM $L_{1.5}$ | | | | | SVM $L_{1.5}$ | | | | |
|---|---|---|---|---|---|---|---|---|---|---|
| | 14p | 15p | 16p | 17p | 18p | 14p | 15p | 16p | 17p | 18p |
| 24g | 0.0552 | **0.0487** | *0.0649* | *0.0779* | *0.0714* | 0.0877 | 0.0747 | 0.0909 | 0.0909 | 0.0942 |
| 25g | 0.0422 | **0.0325** | 0.0487 | 0.0519 | 0.0552 | 0.0682 | 0.0617 | 0.0779 | 0.0747 | 0.0779 |
| 26g | 0.0422 | **0.0357** | 0.0519 | 0.0617 | 0.0617 | 0.0682 | 0.0552 | 0.0714 | 0.0682 | 0.0682 |
| 27g | 0.0390 | 0.0455 | 0.0552 | 0.0552 | *0.0617* | 0.0714 | 0.0617 | 0.0747 | 0.0682 | 0.0682 |
| 28g | **0.0390** | **0.0292** | 0.0455 | 0.0552 | *0.0519* | 0.0682 | 0.0649 | 0.0714 | 0.0682 | 0.0682 |
| | LSSVM $L_{1.3333}$ | | | | | SVM $L_{1.3333}$ | | | | |
| | 14p | 15p | 16p | 17p | 18p | 14p | 15p | 16p | 17p | 18p |
| 24g | 0.0519 | **0.0487** | *0.0649* | *0.0779* | *0.0714* | 0.0812 | 0.0747 | 0.0812 | 0.0844 | 0.0844 |
| 25g | 0.0422 | **0.0325** | *0.0422* | 0.0519 | 0.0552 | 0.0649 | 0.0617 | 0.0747 | 0.0714 | 0.0714 |
| 26g | *0.0390* | **0.0357** | 0.0487 | 0.0584 | *0.0584* | 0.0649 | 0.0552 | 0.0714 | 0.0682 | 0.0682 |
| 27g | 0.0422 | 0.0422 | 0.0552 | 0.0552 | *0.0617* | 0.0682 | 0.0584 | 0.0714 | 0.0682 | 0.0682 |
| 28g | **0.0390** | **0.0292** | **0.0422** | 0.0552 | **0.0487** | 0.0617 | 0.0584 | 0.0682 | 0.0682 | 0.0682 |
| | LSSVM $L_{1.25}$ | | | | | SVM $L_{1.25}$ | | | | |
| | 14p | 15p | 16p | 17p | 18p | 14p | 15p | 16p | 17p | 18p |
| 24g | 0.0519 | **0.0487** | *0.0649* | *0.0779* | **0.0682** | 0.0779 | 0.0649 | 0.0747 | *0.0779* | 0.0812 |
| 25g | **0.0357** | **0.0325** | **0.0390** | *0.0487* | 0.0552 | 0.0649 | 0.0552 | 0.0682 | 0.0714 | 0.0714 |
| 26g | **0.0357** | **0.0357** | *0.0455* | **0.0455** | **0.0455** | 0.0584 | 0.0519 | 0.0682 | 0.0682 | 0.0682 |
| 27g | **0.0357** | *0.0390* | 0.0519 | 0.0552 | *0.0617* | 0.0682 | 0.0584 | 0.0714 | 0.0682 | 0.0682 |
| 28g | **0.0390** | **0.0292** | **0.0422** | *0.0519* | **0.0487** | 0.0617 | 0.0584 | 0.0682 | 0.0682 | 0.0682 |
| | LSSVM $L_{1.2}$ | | | | | SVM $L_{1.2}$ | | | | |
| | 14p | 15p | 16p | 17p | 18p | 14p | 15p | 16p | 17p | 18p |
| 24g | 0.0519 | **0.0487** | **0.0617** | *0.0779* | **0.0682** | 0.0779 | 0.0649 | 0.0747 | *0.0779* | 0.0812 |
| 25g | **0.0357** | **0.0325** | **0.0390** | *0.0487* | 0.0552 | 0.0649 | 0.0552 | 0.0682 | 0.0714 | 0.0714 |
| 26g | **0.0357** | **0.0357** | *0.0455* | 0.0552 | *0.0584* | 0.0649 | 0.0519 | 0.0714 | 0.0682 | 0.0682 |
| 27g | **0.0357** | *0.0390* | *0.0487* | 0.0552 | *0.0617* | 0.0682 | 0.0584 | 0.0714 | 0.0682 | 0.0682 |
| 28g | **0.0390** | **0.0292** | **0.0422** | *0.0519* | **0.0487** | 0.0617 | 0.0584 | 0.0682 | 0.0682 | 0.0682 |
| | LSSVM $L_{1.1667}$ | | | | | SVM $L_{1.1667}$ | | | | |
| | 14p | 15p | 16p | 17p | 18p | 14p | 15p | 16p | 17p | 18p |
| 24g | 0.0519 | **0.0487** | **0.0617** | *0.0779* | **0.0682** | 0.0779 | 0.0617 | 0.0747 | *0.0779* | 0.0812 |
| 25g | **0.0357** | **0.0325** | **0.0390** | *0.0487* | *0.0519* | 0.0649 | 0.0519 | 0.0682 | 0.0714 | 0.0714 |
| 26g | **0.0357** | **0.0357** | **0.0422** | 0.0519 | *0.0584* | 0.0649 | 0.0519 | 0.0714 | 0.0682 | 0.0682 |
| 27g | **0.0357** | *0.0390* | **0.0455** | *0.0519* | *0.0617* | 0.0682 | 0.0584 | 0.0714 | 0.0682 | 0.0682 |
| 28g | **0.0390** | **0.0292** | **0.0422** | *0.0519* | **0.0487** | 0.0617 | 0.0584 | 0.0682 | 0.0682 | 0.0682 |

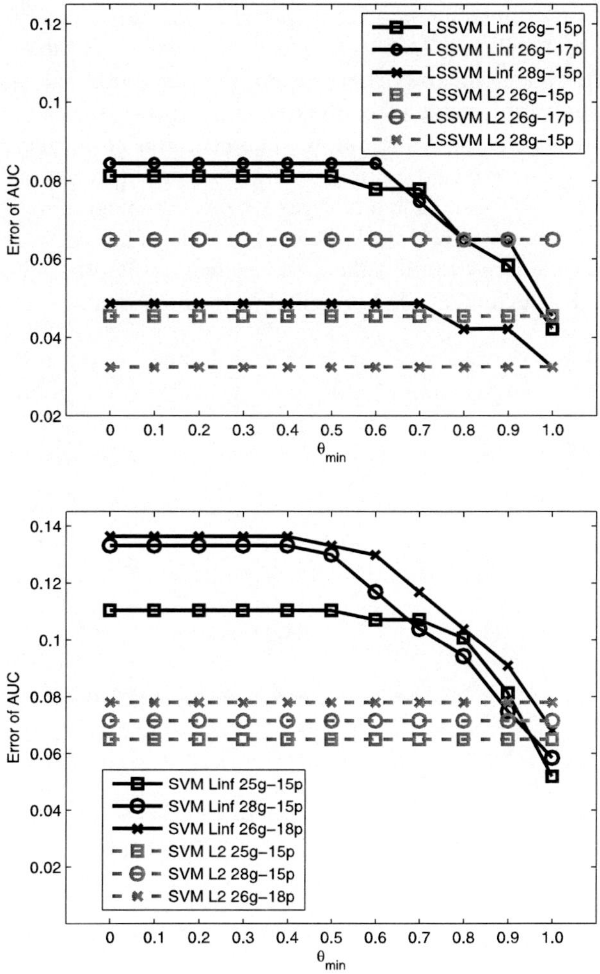


**Fig. 3.6** The effect of $\theta_{min}$ on LSSVM MKL and SVM MKL classifier in rectal cancer diagnosis

In LSSVM kernel fusion, we estimated the $\lambda$ jointly as a coefficient assigned to an identity matrix. Since the number of samples is small in this experiment, the standard cross-validation approach to select the optimal $\lambda$ on validation data was not tried. To investigate whether the estimated $\lambda$ value is optimal, we set $\lambda$ to 51 different values uniformly sampled on the $log_2$ scale from -10 to 40. We compared the joint estimation result with the optimal classification performance among the sampled $\lambda$ values. The joint estimation results were found as optimal for most of the results. An example is illustrated in Figure 3.7 as the integration of four kernels constructed by 27 gene features and 17 protein features. The coefficients estimated by the $L_\infty$-norm were almost 0 thus the $\lambda$ values were very big. In contrast, the $\lambda$ values estimated by the non-sparse $L_2$ method were at reasonable scales.

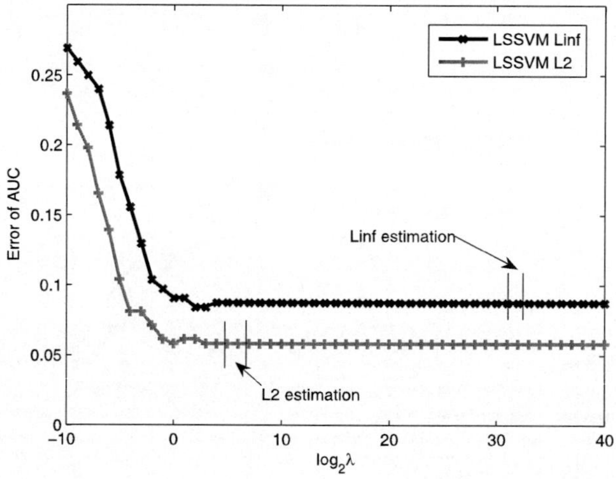


**Fig. 3.7** Benchmark of various $\lambda$ values in LSSVM MKL classifiers on the rectal cancer diagnosis problem

## Experiment 4: Clinical Decision Support by Integrating Multiple Kernels

In the fourth experiment we validated the proposed approach on three clinical data sets containing more samples. On the endometrial and miscarriage data sets, we compared eight MKL algorithms with various norms. For the imbalanced pregnancy data set, we applied eight weighted MKL algorithms. The results are shown in Table 3.12, 3.13, and 3.14. On endometrial data, the difference of performance was rather small. Though the two $L_2$ methods were not optimal, they were comparable to the best result. On miscarriage data, the $L_2$ methods performed significantly better than comparing algorithms. On pregnancy data, the weighted $L_2$ LSSVM MKL and weighted $L_1$ LSSVM MKL performed significantly better than others.

**Table 3.12** Results of experiment 4 data set I: classification of endometrial disease patients using multiple kernels derived from clinical data. The classifier with the best performance is shown in bold. The p-values are compared with the best performance using a paired t-test. The performance of $L_\infty$, $L_1$, and $L_2$ MKL classifiers is sorted from high to low according to the p-values.

| Classifier | Mean - error of AUC | Std. - error of AUC | pvalue |
|---|---|---|---|
| **LSSVM $L_\infty$ (0.5) MKL** | **0.2353** | **0.0133** | - |
| **SVM $L_\infty$ (0.5) MKL** | **0.2388** | **0.0178** | 0.4369 |
| **SVM $L_\infty$ MKL** | **0.2417** | **0.0165** | 0.2483 |
| LSSVM $L_2$ MKL | 0.2456 | 0.0124 | 0.0363 |
| SVM $L_2$ MKL | 0.2489 | 0.0178 | 0.0130 |
| SVM $L_1$ MKL | 0.2513 | 0.0144 | 0.0057 |
| LSSVM $L_1$ MKL | 0.2574 | 0.0189 | $9.98 \cdot 10^{-5}$ |
| LSSVM $L_\infty$ MKL | 0.2678 | 0.0130 | $1.53 \cdot 10^{-6}$ |
| LSSVM $L_{1.5}$ MKL | 0.2427 | 0.0107 | |
| LSSVM $L_{1.3333}$ MKL | 0.2446 | 0.0100 | |
| LSSVM $L_{1.25}$ MKL | 0.2466 | 0.0114 | |
| LSSVM $L_{1.2}$ MKL | 0.2475 | 0.0115 | |
| LSSVM $L_{1.1667}$ MKL | 0.2477 | 0.0142 | |
| SVM $L_{1.5}$ MKL | 0.2360 | 0.0082 | |
| SVM $L_{1.3333}$ MKL | 0.2375 | 0.0081 | |
| SVM $L_{1.25}$ MKL | 0.2373 | 0.0085 | |
| SVM $L_{1.2}$ MKL | 0.2368 | 0.0086 | |
| SVM $L_{1.1667}$ MKL | 0.2369 | 0.0089 | |

**Table 3.13** Results of experiment 4 data set II: classification of miscarriage patients using multiple kernels derived from clinical data. The classifier with the best performance is shown in bold. The p-values are compared with the best performance using a paired t-test. The performance of $L_\infty$, $L_1$, and $L_2$ MKL classifiers is sorted from high to low according to the p-values.

| Classifier | Mean - error of AUC | Std. - error of AUC | pvalue |
|---|---|---|---|
| **SVM $L_2$ MKL** | **0.1975** | **0.0037** | - |
| **LSSVM $L_2$ MKL** | **0.2002** | **0.0049** | 0.0712 |
| LSSVM $L_\infty$ (0.5) MKL | 0.2027 | 0.0045 | $9.77 \cdot 10^{-4}$ |
| SVM $L_\infty$ MKL | 0.2109 | 0.0040 | $9.55 \cdot 10^{-12}$ |
| SVM $L_\infty$ (0.5) MKL | 0.2168 | 0.0040 | $1.79 \cdot 10^{-12}$ |
| LSSVM $L_1$ MKL | 0.2132 | 0.0029 | $1.11 \cdot 10^{-13}$ |
| SVM $L_1$ MKL | 0.2297 | 0.0038 | $1.10 \cdot 10^{-15}$ |
| LSSVM $L_\infty$ MKL | 0.2319 | 0.0015 | $3.42 \cdot 10^{-21}$ |
| LSSVM $L_{1.5}$ MKL | 0.1892 | 0.0081 | |
| LSSVM $L_{1.3333}$ MKL | 0.1921 | 0.0096 | |
| LSSVM $L_{1.25}$ MKL | 0.1906 | 0.0074 | |
| LSSVM $L_{1.2}$ MKL | 0.1927 | 0.0080 | |
| LSSVM $L_{1.1667}$ MKL | 0.1882 | 0.0064 | |
| SVM $L_{1.5}$ MKL | 0.2116 | 0.0050 | |
| SVM $L_{1.3333}$ MKL | 0.2102 | 0.0042 | |
| SVM $L_{1.25}$ MKL | 0.2091 | 0.0056 | |
| SVM $L_{1.2}$ MKL | 0.2077 | 0.0038 | |
| SVM $L_{1.1667}$ MKL | 0.2093 | 0.0040 | |

**Table 3.14** Results of experiment 4 data set III: classification of PUL patients using multiple kernels derived from clinical data. The classifier with the best performance is shown in bold. The p-values are compared with the best performance using a paired t-test. The performance of classifiers is sorted from high to low according to the p-values.

| Classifier | Mean - error of AUC | Std. - error of AUC | pvalue |
|---|---|---|---|
| **Weighted LSSVM $L_2$ MKL** | **0.1165** | **0.0100** | - |
| **Weighted LSSVM $L_1$ MKL** | **0.1243** | **0.0171** | 0.0519 |
| Weighted LSSVM $L_\infty$ (0.5) MKL | 0.1290 | 0.0206 | 0.0169 |
| Weighted SVM $L_2$ MKL | 0.1499 | 0.0248 | $4.79 \cdot 10^{-5}$ |
| Weighted SVM $L_\infty$ MKL | 0.1552 | 0.0210 | $1.02 \cdot 10^{-6}$ |
| Weighted SVM $L_\infty$ (0.5) | 0.1551 | 0.0153 | $3.87 \cdot 10^{-6}$ |
| Weighted SVM $L_1$ MKL | 0.1594 | 0.0162 | $2.29 \cdot 10^{-9}$ |
| Weighted LSSVM $L_\infty$ MKL | 0.1651 | 0.0174 | $4.41 \cdot 10^{-10}$ |
| Weighted LSSVM $L_{1.5}$ MKL | 0.1086 | 0.0067 | |
| Weighted LSSVM $L_{1.3333}$ MKL | 0.1076 | 0.0069 | |
| Weighted LSSVM $L_{1.25}$ MKL | 0.1068 | 0.0070 | |
| Weighted LSSVM $L_{1.2}$ MKL | 0.1112 | 0.0129 | |
| Weighted LSSVM $L_{1.1667}$ MKL | 0.1099 | 0.0100 | |
| Weighted SVM $L_{1.5}$ MKL | 0.1244 | 0.0152 | |
| Weighted SVM $L_{1.3333}$ MKL | 0.1213 | 0.0107 | |
| Weighted SVM $L_{1.25}$ MKL | 0.1234 | 0.0109 | |
| Weighted SVM $L_{1.2}$ MKL | 0.1228 | 0.0141 | |
| Weighted SVM $L_{1.1667}$ MKL | 0.1199 | 0.0137 | |

To investigate whether the combination of multiple kernels performs as well as the best individual kernel, we evaluated the performance of all the individual kernels in Table 3.15. As shown, the clinical kernel proposed by Daemen and De Moor [14] has better quality than linear, RBF and polynomial kernels on endometrial and pregnancy data sets. For the miscarriage data set, the first RBF kernel has better quality than the other seven kernels. Despite the difference in individual kernels, the performance of MKL is comparable to the best individual kernel, demonstrating that MKL is also useful to combine candidate kernels derived from a single data set.

**Table 3.15** Performance of individual kernels in Experiment 4. For each combination of data set and algorithm, the best individual kernels are shown in bold. For each data set across different single kernel algorithms, the best results are underlined. The best MKL performance is also shown for comparison. Obviously, MKL performance is comparable to the results of best individual kernels.

| Data Set | Classifier | Kernel | Error of AUC (mean) | Error of AUC (std.) |
|---|---|---|---|---|
| endometrial | LSSVM | linear | 0.2820 | 0.0175 |
| | | RBF1 | 0.2923 | 0.0131 |
| | | RBF2 | 0.2844 | 0.0118 |
| | | RBF3 | 0.2915 | 0.0119 |
| | | POLY1 | 0.3223 | 0.0088 |
| | | POLY2 | 0.3226 | 0.0109 |
| | | POLY3 | 0.3183 | 0.0128 |
| | | Clinical | **0.2126** | **0.0098** |
| | SVM | linear | 0.2816 | 0.0192 |
| | | RBF1 | 0.2971 | 0.0112 |
| | | RBF2 | 0.2817 | 0.0098 |
| | | RBF3 | 0.2877 | 0.0133 |
| | | POLY1 | 0.3271 | 0.0141 |
| | | POLY2 | 0.3214 | 0.0130 |
| | | POLY3 | 0.3225 | 0.0135 |
| | | Clinical | **0.2021** | **0.0084** |
| | best MKL classifier | | 0.2353 | 0.0133 |
| miscarriage | LSSVM | linear | 0.2410 | 0.0022 |
| | | RBF1 | **0.1993** | **0.0042** |
| | | RBF2 | 0.2114 | 0.0029 |
| | | RBF3 | 0.2182 | 0.0030 |
| | | POLY1 | 0.2637 | 0.0020 |
| | | POLY2 | 0.2607 | 0.0023 |
| | | POLY3 | 0.2593 | 0.0019 |
| | | Clinical | 0.2301 | 0.0026 |
| | SVM | linear | 0.2781 | 0.0065 |
| | | RBF1 | **0.2098** | **0.0029** |
| | | RBF2 | 0.2272 | 0.0042 |
| | | RBF3 | 0.2352 | 0.0037 |
| | | POLY1 | 0.2771 | 0.0013 |
| | | POLY2 | 0.2741 | 0.0023 |
| | | POLY3 | 0.2713 | 0.0016 |
| | | Clinical | 0.2441 | 0.0035 |
| | best MKL classifier | | 0.1892 | 0.0081 |
| pregnancy | Weighted LSSVM | linear | 0.1666 | 0.0118 |
| | | RBF1 | 0.1763 | 0.0142 |
| | | RBF2 | 0.1990 | 0.0146 |
| | | RBF3 | 0.2137 | 0.0170 |
| | | POLY1 | 0.2836 | 0.0154 |
| | | POLY2 | 0.2639 | 0.0169 |
| | | POLY3 | 0.2382 | 0.0180 |
| | | Clinical | **0.1160** | **0.0092** |
| | Weighted SVM | linear | **0.1461** | **0.0074** |
| | | RBF1 | 0.2127 | 0.0187 |
| | | RBF2 | 0.2017 | 0.0254 |
| | | RBF3 | 0.1906 | 0.0221 |
| | | POLY1 | 0.1478 | 0.0184 |
| | | POLY2 | 0.1541 | 0.0204 |
| | | POLY3 | 0.1594 | 0.0179 |
| | | Clinical | 0.1601 | 0.0188 |
| | best MKL classifier | | 0.1068 | 0.0070 |

In the results presented before, the regularization parameter $\lambda$ in LSSVM classifiers was jointly estimated in MKL. Since the clinical data sets contain a sufficient number of samples to select the $\lambda$ by cross validation, we systematically compared the estimation approach with the standard validation approach to determine the $\lambda$ values. As shown in Table 3.16, the estimation approach based on $L_\infty$ performed worse than the validation approach. This is probably because the estimated $\lambda$ values are either very big or very small when the kernel coefficients were sparse. In contrast, the $L_2$ based estimation approach yielded comparable performance as the validation approach. We also benchmarked the performance of LSSVM MKL classifiers using 21 different static $\lambda$ values on the data sets and the results are shown in Figure 3.8. In real problems, to select the optimal $\lambda$ value in LSSVM is a non-trivial issue and it is often optimized as a hyper-parameter on validation data. The main advantage of $L_2$ and $L_n$ MKL is that the estimation approach is more computational efficient than cross validation and yields a comparable performance.

**Table 3.16** Comparison of the performance obtained by joint estimation of $\lambda$ and standard cross-validation in LSSVM MKL

| Data Set | Norm | Validation Approach | Estimation Approach |
|---|---|---|---|
| endometrial disease | $L_\infty$ | $0.2625 \pm 0.0146$ | $0.2678 \pm 0.0130$ |
| | $L_2$ | $0.2584 \pm 0.0188$ | $0.2456 \pm 0.0124$ |
| miscarriage | $L_\infty$ | $0.1873 \pm 0.0100$ | $0.2319 \pm 0.0015$ |
| | $L_2$ | $0.1912 \pm 0.0089$ | $0.2002 \pm 0.0049$ |
| pregnancy | $L_\infty$ | $0.1321 \pm 0.0243$ | $0.1651 \pm 0.0173$ |
| | $L_2$ | $0.1299 \pm 0.0172$ | $0.1165 \pm 0.0100$ |

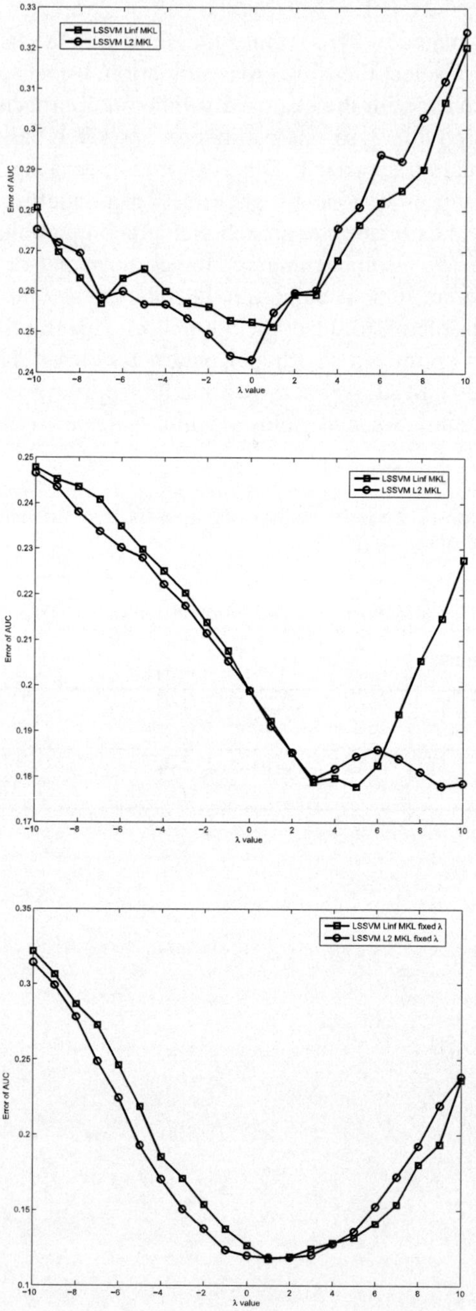


**Fig. 3.8** The performance of LSSVM MKL classifiers varied by various $\lambda$ values on endometrial (top), miscarriage (middle), and pregnancy (bottom) disease data set

## 3.11 Discussions

In this chapter we proposed a new $L_2$ MKL framework as the complement to the existing $L_\infty$ MKL method proposed by Lanckriet *et al.*. The $L_2$ MKL is characterized by the non-sparse integration of multiple kernels to optimize the objective function of machine learning problems. On four real bioinformatics and biomedical applications, we systematically validated the performance through extensive analysis. The motivation for $L_2$ MKL is as follows. In real biomedical applications, with a small number of sources that are believed to be truly informative, we would usually prefer a nonsparse set of coefficients because we would want to avoid that the dominant source (like text mining or Gene Ontology) gets a coefficient close to 1. The reason to avoid sparse coefficients is that there is a discrepancy between the experimental setup for performance evaluation and "real world" performance. The dominant source will work well on a benchmark because this is a controlled situation with known outcomes. We for example set up a set of already known genes for a given disease and want to demonstrate that our model can capture the available information to discriminate between a gene from this set and randomly selected genes (for example, in a cross-validation setup). Given that these genes are already known to be associated with the disease, this information will be present in sources like text mining or Gene Ontology in the gene prioritization problem. These sources can then identify these known genes with high confidence and should therefore be assigned a high weight. However, when trying to identify truly novel genes for the same disease, the relevance of the information available through such data sources will be much lower and we would like to avoid anyone data source to complete dominate the other. Given that setting up a benchmark requires knowledge of the association between a gene and a disease, this effect is hard to avoid. We can therefore expect that if we have a smoother solution that performs as well as the sparse solution on benchmark data, it is likely to perform better on real discoveries.

For the specific problem of gene prioritization, an effective way to address this problem is to setup a benchmark where information is "rolled back" a number of years (*e.g.*, two years) prior to the discovery of the association between a gene and a disease (i.e., older information is used so that the information about the association between the gene and the disease is not yet contained in data sources like text mining or Gene Ontology). Given that the date at which the association was discovered is different for each gene, the setup of such benchmarks is notoriously difficult. In future work, we plan to address this problem by freezing available knowledge at a given data and then collecting novel discoveries and benchmarking against such discoveries in a fashion reminiscent of CASP (Critical Assessment of protein Structure Prediction) [33].

The technical merit of the proposed $L_2$ MKL lies in the dual forms of various objective functions. Though in the literature the issue of using different norms in MKL is recently investigated by Kloft *et al.* [26, 27] and Kowalski *et al.* [28], their formulations are based on the primal problems. We have theoretically proven that optimizing the $L_2$ regularization of kernel coefficients in the primal problem corresponds to solving the $L_2$-norm of kernel components in the dual problem. Clarifying this dual solution enabled us to directly solve the $L_2$ problem as a convex SOCP. Moreover, the dual solution can be extended to various other machine learning problems. In this paper we have shown the extensions of 1-SVM, SVM and LSSVM. As a matter of fact, the $L_2$ dual solution can also be applied in kernel based clustering analysis and regression analysis for a wide range of applications.

Another main contribution of our paper is the novel LSSVM $L_2$ MKL proposed for classification problems. As known, when applying various machine learning techniques to solve real computational biological problems, the performance may depend on the data set and the experimental settings. When the performance evaluations of various methods are comparable, but with one method showing significant computational efficiency over other methods, this would be a "solid" advantage of this method. In this paper, we have shown that the LSSVM MKL classifier based on SIP formulation can be solved more efficiently than SVM MKL. Moreover, the performance of LSSVM $L_2$ MKL is always comparable to the best performance. The SIP based LSSVM $L_2$ MKL classifier has two main "solid advantages": the inherent time complexity is small and the regularization parameter $\lambda$ can be jointly estimated in the experimental setup. Due to these merits, LSSVM $L_2$ MKL is a very promising technique for problems pertaining to large scale data fusion.

## 3.12 Summary

In this chapter, we compared the effect of optimizing different norms in multiple kernel learning in a systematic framework. The obtained results extend and enrich the statistical framework of genomic data fusion proposed by Lanckriet *et al.* [29, 30] and Bach *et al.* [6]. According to the optimization of different norms in the dual problem of SVM, we proposed $L_\infty$, $L_1$, $L_2$, and $L_n$ MKL, which are respectively corresponding to the $L_1$ regularization, average combination, $L_2$, and $L_m$ regularization of kernel coefficients addressed in the primal problem. We have proved that $n = \frac{m}{m-1}$.

Six real biomedical data sets were investigated in this paper, where $L_2$ MKL approach was shown advantageous over the $L_\infty$ method. We also proposed a novel and efficient LSSVM $L_2$ MKL classifier to learn the optimal combination of multiple large scale data sets. All the algorithms implemented in this paper are freely accessible on http://homes.esat.kuleuven.be/∼sistawww/bioi/syu/l2lssvm.html.

## References


1. Aerts, S., Lambrechts, D., Maity, S., Van Loo, P., Coessens, B., De Smet, F., Tranchevent, L.C., De Moor, B., Marynen, P., Hassan, B., Carmeliet, P., Moreau, Y.: Gene prioritization through genomic data fusion. Nature Biotechnology 24, 537–544 (2006)
2. Aerts, S., Van Loo, P., Thijs, G., Mayer, H., de Martin, R., Moreau, Y., De Moor, B.: TOUCAN 2: the all-inclusive open source workbench for regulatory sequence analysis. Nucleic Acids Research 396, W393–W396 (2005)
3. Aizerman, M., Braverman, E., Rozonoer, L.: Theoretical foundations of the potential function method in pattern recognition learning. Automation and Remote Control 25, 821–837 (1964)
4. Andersen, E.D., Andersen, K.D.: The MOSEK interior point optimizer for linear programming: an implementation of the homogeneous algorithm. In: High Perf. Optimization, pp. 197–232. Kluwer Academic Publishers, New York (2000)

5. Ashburner, M., Ball, C.A., Blake, J.A., Botstein, D., Butler, H., Cherry, J.M., Davis, A.P., Dolinski, K., Dwight, S.S., Eppig, J.T., Harris, M.A., Hill, D.P., Issel-Tarver, L., Kasarskis, A., Lewis, S., Matese, J.C., Richardson, J.E., Ringwald, M., Rubin, G.M., Sherlock, G.: Gene ontology: tool for the unification of biology. The Gene Ontology Consortium. Nature Genetics. 25, 25–29 (2000)
6. Bach, F.R., Lanckriet, G.R.G., Jordan, M.I.: Multiple kernel learning, conic duality, and the SMO algorithm. In: Proceedings of 21st International Conference of Machine Learning. ACM Press, New York (2004)
7. van den Bosch, T., Daemen, A., Gevaert, O., Timmerman, D.: Mathematical decision trees versus clinician based algorithms in the diagnosis of endometrial disease. In: Proc. of the 17th World Congress on Ultrasound in Obstetrics and Gynecology (ISUOG), vol. 412 (2007)
8. Boser, B.E., Guyon, I.M., Vapnik, V.N.: A training algorithm for optimal margin classifiers. In: Proceedings of the 5th Annual ACM Workshop on COLT, pp. 144–152. ACM Press, New York (1992)
9. Bottomley, C., Daemen, A., Mukri, F., Papageorghiou, A.T., Kirk, E., Pexsters, A., De Moor, B., Timmerman, D., Bourne, T.: Functional linear discriminant analysis: a new longitudinal approach to the assessment of embryonic growth. Human Reproduction 24, 278–283 (2007)
10. Boyd, S., Vandenberghe, L.: Convex Optimization. Cambridge University Press, Cambridge (2004)
11. Cawley, G.C.: Leave-One-Out Cross-Validation Based Model Selection Criteria for Weighted LS-SVMs. In: Proc. of 2006 International Joint Conference on Neural Networks, pp. 1661–1668. IEEE press, Los Alamitos (2006)
12. Condous, G., Okaro, E., Khalid, A., Timmerman, D., Lu, C., Zhou, Y., Van Huffel, S., Bourne, T.: The use of a new logistic regression model for predicting the outcome of pregnancies of unknown location. Human Reproduction 21, 278–283 (2004)
13. Daemen, A., De Moor, B.: Development of a kernel function for clinical data. In: Proc. of the 31th Annual International Conference of the IEEE Engineering in Medicine and Biology Society (EMBC), pp. 5913–5917. IEEE press, Los Alamitos (2009)
14. Daemen, A., Gevaert, O., Ojeda, F., Debucquoy, A., Suykens, J.A.K., Sempous, C., Machiels, J.P., Haustermans, K., De Moor, B.: A kernel-based integration of genome-wide data for clinical decision support. Genome Medicine 1, 39 (2009)
15. De Bie, T., Tranchevent, L.C., Van Oeffelen, L., Moreau, Y.: Kernel-based data fusion for gene prioritization. Bioinformatics 132, i125–i132 (2007)
16. Eeles, R.A., Kote-Jarai, Z., Giles, G.G., Olama, A.A.A., Guy, M., Jugurnauth, S.K., Mulholland, S., Leongamornlert, D.A., Edwards, S.M., Morrison, J., et al.: Multiple newly identified loci associated with prostate cancer susceptibility. Nature Genetics 40, 316–321 (2008)
17. Flicek, P., Aken, B.L., Beal, K., Ballester, B., Caccamo, M., Chen, Y., Clarke, L., Caotes, G., Gunningham, F., Cutts, T., Down, T., Dyer, S.C., Eyre, T., Fitzgerald, S., Fernandez-Banet, J., Gräf, S., Haider, S., Hammond, R., Holland, R., Howe, K.L., Howe, K., Johnson, N., Jenkinson, A., Kähäri, A., Keefe, D., Kokocinski, F., Kulesha, E., Lawson, D., Longden, I., Megy, K., Meidl, P., Overduin, B., Parker, A., Pritchard, B., Prlic, A., Rice, S., Rios, D., Schuster, M., Sealy, I., Slater, G., Smedley, D., Spudich, G., Trevanion, S., Vilella, A.J., Vogel, J., White, S., Wood, M., Birney, E., Cox, T., Curwen, V., Durbin, R., Fernandez-Suarez, X.M., Herrero, J., Hubbard, T.J.P., Kasprzyk, A., Proctor, G., Smith, J., Ureta-Vidal, A., Searle, S.: Ensembl 2008. Nucleic Acids Research 36, D707–D714 (2007)

18. Gevaert, O., De Smet, F., Timmerman, D., Moreau, Y., De Moor, B.: Predicting the prognosis of breast cancer by integrating clinical and microarray data with Bayesian networks. Bioinformatics 190, e184–e190 (2006)
19. Grant, M., Boyd, S.: Graph implementations for nonsmooth convex programs. Recent Advances in Learning and Control 371, 95–110 (2008)
20. Grant, M., Boyd, S.: CVX: Matlab Software for Disciplined Convex Programming, version 1.21 (2010), `http://cvxr.com/cvx`
21. Gudmundsson, J., Sulem, P., Rafnar, T., Bergthorsson, J.T., Manolescu, A., Gudbjartsson, D., Agnarsson, B.A., Sigurdsson, A., Benediktsdottir, K.R., Blondal, T., et al.: Common sequence variants on 2p15 and Xp11.22 confer susceptibility to prostate cancer. Nature Genetics 40, 281–283 (2008)
22. Hettich, R., Kortanek, K.O.: Semi-infinite programming: theory, methods, and applications. SIAM Review 35, 380–429 (1993)
23. Kaliski, J., Haglin, D., Roos, C., Terlaky, T.: Logarithmic barrier decomposition methods for semi-infinite programming. International Transactions in Operations Research 4, 285–303 (1997)
24. Kanehisa, M., Araki, M., Goto, S., Hattori, M., Hirakawa, M., Itoh, M., Katayama, T., Kawashima, S., Okuda, S., Tokimatsu, T., Yamanishi, Y.: KEGG for linking genomes to life and the environment. Nucleic Acids Research 36, D480–D484 (2008)
25. Kim, S.J., Magnani, A., Boyd, S.: Optimal kernel selection in kernel fisher discriminant analysis. In: Proceeding of 23rd International Conference of Machine Learning. ACM Press, New York (2006)
26. Kloft, M., Brefeld, U., Laskov, P., Sonnenburg, S.: Non-sparse multiple kernel learning. In: NIPS 2008 Workshop: Kernel Learning Automatic Selection of Optimal Kernels (2008)
27. Kloft, M., Brefeld, U., Sonnenburg, S., Laskov, P., Müller, K.R., Zien, A.: Efficient and Accurate Lp-norm Multiple Kernel Learning. In: Advances in Neural Information Processing Systems, vol. 22. MIT Press, Cambridge (2009)
28. Kowalski, M., Szafranski, M., Ralaivola, L.: Multiple indefinite kernel learning with mixed norm regularization. In: Proc. of the 26th International Conference of Machine Learning. ACM Press, New York (2009)
29. Lanckriet, G.R.G., Cristianini, N., Bartlett, P., Ghaoui, L.E., Jordan, M.I.: Learning the Kernel Matrix with Semidefinite Programming. Journal of Machine Learning Research 5, 27–72 (2005)
30. Lanckriet, G.R.G., De Bie, T., Cristianini, N., Jordan, M.I., Noble, W.S.: A statistical framework for genomic data fusion. Bioinformatics 20, 2626–2635 (2004)
31. Leslie, C., Eskin, E., Weston, J., Noble, W.S.: The spectrum kernel: a string kernel for SVM protein classification. In: Proc. of the Pacific Symposium on Biocomputing 2002, pp. 564–575 (2002)
32. Matys, V., Fricke, E., Geffers, R., Gößling, E., Haubrock, M., Hehl, R., Hornischer, K., Karas, D., Kel, A.E., Kel-Margoulis, O.V., Kloos, D.-U., Land, S., Lewicki-Potapov, B., Michael, H., Münch, R., Reuter, I., Rotert, S., Saxel, H., Scheer, M., Thiele, S., Wingender, E.: TRANSFAC: transcriptional regulation, from patterns to profiles. Nucleic Acids Research 31, 374–378 (2003)
33. Moult, J., Fidelis, K., Kryshtafovych, A., Rost, B., Tramontano, A.: Critical assessment of methods of protein structure prediction - Round VIII. Proteins 69(S8), 3–9 (2009)

34. Mulder, N.J., Apweiler, R., Attwood, T.K., Bairoch, A., Bateman, A., Binns, D., Bork, P., Buillard, V., Cerutti, L., Copley, R., Courcelle, E., Das, U., Daugherty, L., Dibley, M., Finn, R., Fleischmann, W., Gough, J., Haft, D., Hulo, N., Hunter, S., Kahn, D., Kanapin, A., Kejariwal, A., Labarga, A., Langendijk-Genevaux, P.S., Lonsdale, D., Lopez, R., Letunic, I., Madera, M., Maslen, J., McAnulla, C., McDowall, J., Mistry, J., Mitchell, A., Nikolskaya, A.N., Orchard, S., Orengo, C., Petryszak, R., Selengut, J.D., Sigrist, C.J.A., Thomas, P.D., Valentin, F., Wilson, D., Wu, C.H., Yeats, C.: New developments in the InterPro database. Nucleic Acids Research 35, D224–D228 (2007)
35. Ng, A.Y.: Feature selection, L1 vs. L2 regularization, and rotational invariance. In: Proceedings of 21st International Conference of Machine Learning. ACM Press, New York (2004)
36. Osuna, E., Freund, R., Girosi, F.: Support vector machines: Training and applications. Tech. Rep. AIM-1602 (1997)
37. Reemtsen, R.: Some other approximation methods for semi-infinite optimization problems. Jounral of Computational and Applied Mathematics 53, 87–108 (1994)
38. Schölkopf, B., Platt, J.C., Shawe-Taylor, J., Smola, A.J., Williamson, R.C.: Estimating the support of a high-dimensional distribution. Neural Computation 13, 1443–1471 (2001)
39. Sturm, J.F.: Using SeDuMi 1.02, a MATLAB toolbox for optimization over symmetric cones. Optimization Methods and Software 11/12, 625–653 (1999)
40. Shawe-Taylor, J., Cristianini, N.: Kernel methods for pattern analysis. Cambridge University Press, Cambridge (2004)
41. Son, C.G., Bilke, S., Davis, S., Greer, B.T., Wei, J.S., Whiteford, C.C., Chen, Q.R., Cenacchi, N., Khan, J.: Database of mRNA gene expression profiles of multiple human organs. Genome Research 15, 443–450 (2005)
42. Sonnenburg, S., Rätsch, G., Schäfer, C., Schölkopf, B.: Large scale multiple kernel learning. Journal of Machine Learning Research 7, 1531–1565 (2006)
43. Su, A.I., Cooke, M.P., Ching, K.A., Hakak, Y., Walker, J., Wiltshire, T., Orth, A.P., Vega, R.G., Sapinoso, L.M., Moqrich, A., Patapoutian, A., Hampton, G.M., Schultz, P.G., Hogenesch, J.B.: Large-scale analysis of the human and mouse transcriptomes. PNAS 99, 4465–4470 (2002)
44. Suykens, J.A.K., De Brabanter, J., Lukas, L., Vandewalle, J.: Weighted least squares support vector machines: robustness and sparse approximation. Neurocomputing, Special issue on fundamental and information processing aspects of neurocomputing 48, 85–105 (2002)
45. Suykens, J.A.K., Van Gestel, T., Brabanter, J., De Moor, B., Vandewalle, J.: Least Squares Support Vector Machines. World Scientific Press, Singapore (2002)
46. Suykens, J.A.K., Vandewalle, J.: Multiclass Least Squares Support Vector Machines. In: Proc. of IJCNN 1999. IEEE, Los Alamitos (1999)
47. Suykens, J.A.K., Vandewalle, J.: Least squares support vector machine classifiers. Neural Processing Letters 9, 293–300 (1999)
48. Tax, D.M.J., Duin, R.P.W.: Support vector domain description. Pattern Recognition Letter 20, 1191–1199 (1999)
49. Thomas, G., Jacobs, K.B., Yeager, M., Kraft, P., Wacholder, S., Orr, N., Yu, K., Chatterjee, N., Welch, R., Hutchinson, A., et al.: Multiple loci identified in a genome-wide association study of prostate cancer. Nature Genetics 40, 310–315 (2008)
50. Tretyakov, K.: Methods of genomic data fusion: An overview. Internal Report, Institute of Computer Science, University of Tartu (2006)

51. Vapnik, V.: The Nature of Statistical Learning Theory, 2nd edn. Springer, New York (1999)
52. Veropoulos, K., Cristianini, N., Campbell, C.: Controlling the sensitivity of support vector machines. In: Proc. of the IJCAI 1999, pp. 55–60. Morgan Kaufmann Press, San Francisco (1999)
53. Ye, J., McGinnis, S., Madden, T.L.: BLAST: improvements for better sequence analysis. Nucleic Acids Research 34, W6–W9 (2006)
54. Ye, J.P., Ji, S.H., Chen, J.H.: Multi-class discriminant kernel learning via convex programming. Journal of Machine Learning Research 40, 719–758 (2008)
55. Yu, S., Tranchevent, L.-C., De Moor, B., Moreau, Y.: Gene prioritization and clustering by multi-view text mining. BMC Bioinformatics 11, 1–48 (2010)
56. Yu, S., Tranchevent, L.-C., Liu, X., Glänzel, W., Suykens, J.A.K., De Moor, B., Moreau, Y.: Optimized data fusion for kernel K-means clustering. Internal Report, K.U.Leuven (2008) (submitted for publication)
57. Yu, S., Van Vooren, S., Tranchevent, L.-C., De Moor, B., Moreau, Y.: Comparison of vocabularies, representations and ranking algorithms for gene prioritization by text mining. Bioinformatics 24, i119–i125 (2008)
58. Yu, S., Tranchevent, L.-C., Liu, X., Glänzel, W., Suykens, J.A.K., De Moor, B., Moreau, Y.: Optimized data fusion for kernel K-means clustering. Internal Report 08-200, ESAT-SISTA, K.U.Leuven, Lirias number: 242275 (2008) (submitted for publication)
59. Zheng, Y., Yang, X., Beddoe, G.: Reduction of False Positives in Polyp Detection Using Weighted Support Vector Machines. In: Proc. of the 29th Annual International Conference of the IEEE Engineering in Medicine and Biology Society (EMBC), pp. 4433–4436. IEEE Press, Los Alamitos (2007)

# Chapter 4
# Optimized Data Fusion for Kernel $k$-means Clustering

## 4.1 Introduction

In this chapter, we will present a novel *optimized kernel k-means clustering* (OKKC) algorithm to combine multiple data sources. The objective of $k$-means clustering is formulated as a Rayleigh quotient function of the between-cluster scatter and the cluster membership matrix. To incorporate multiple data sources, the between-cluster matrix is calculated in the high dimensional Hilbert space where the heterogeneous data sources can be easily combined as kernel matrices. The objective to optimize the kernel combination and the cluster memberships on unlabeled data is non-convex. To solve it, we apply an alternating minimization [6] method to optimize the cluster memberships and the kernel coefficients iteratively to convergence. When the cluster membership is given, we optimize the kernel coefficients as kernel Fisher Discriminant (KFD) and solve it as least squares support vector machine (LSSVM). The objectives of KFD and $k$-means are combined in a unified model thus the two components optimize towards the same objective, therefore, the proposed alternating algorithm converges locally.

Our algorithm is based on the same motivation as Lange and Buhmann's approach [17] to combine multiple information sources as similarity matrices (kernel matrices). However, the two algorithmic approaches are different. Lange and Buhmann's algorithm uses non-negative matrix factorization to maximize posteriori estimates to obtain the assignment of data points to partitions. To fuse the similarity matrices, they minimize the cross-entropy to seek a good factorization and the optimal weights assigned on similarity matrices. In our approach, the objective is extended from $k$-means clustering and extended as dual representations to combine multiple data sources. The cluster assignments of data points are relaxed as numerical values and optimized as the eigenspectrum of the combined kernel matrix. The coefficients of kernel matrices are optimized as a dual problem of the objective function.

The proposed algorithm is related to the Nonlinear Adaptive Metric Learning (NAML) algorithm proposed for clustering [5], however, it has much simpler structure. Though NAML is also based on multiple kernel extension of $k$-means

S. Yu et al.: Kernel-based Data Fusion for Machine Learning, SCI 345, pp. 89–107.
springerlink.com 

clustering, the mathematical objective and the solution are different from OKKC. In NAML, the metric of $k$-means is constructed based on the Mahalanobis distance and then extended to Hilbert space using the Representer Theorem [22]. NAML optimizes the objective iteratively at three levels: the cluster assignments, the kernel coefficients and the projection in the Representer Theorem. In contrast, our proposed method only optimizes the cluster assignments and kernel coefficients in a bi-level procedure, which is simpler and more efficient than NAML. Moreover, we formulate the least squares dual problem of kernel coefficient learning as semi-infinite programming (SIP) [28], which is much more efficient and scalable than the quadratically constrained quadratic programming (QCQP) [4] formulation adopted in NAML.

This chapter is organized as follows. Section 4.2 introduces the objective of $k$-means clustering. Section 4.3 presents the problem of $k$-means clustering in Hilbert space and the extension to combine multiple data sources. In Section 4.4, we introduce the proposed algorithm to solve the objective. The description of experimental data and analysis of results are presented in Section 4.5. The final summary and conclusion can be found in Section 4.6.

## 4.2 Objective of $k$-means Clustering

In $k$-means clustering, a number of $k$ prototypes are used to characterize the data and the partitions $\{\mathscr{C}_j\}_{j=1,\dots,k}$ are determined by minimizing the distortion as

$$\text{minimize} \sum_{j=1}^{k} \sum_{\mathbf{x}_i \in \mathscr{C}_j} ||\mathbf{x}_i - \mu_j||^2, \tag{4.1}$$

where $\mathbf{x}_i$ is the $i$-th data sample, $\mu_j$ is the prototype (mean) of the $j$-th partition $\mathscr{C}_j$, $k$ is the number of partitions (usually predefined). It is known that (4.1) is equal to the trace maximization of the between-cluster scatter $S_b$ [15, 29], given by

$$\underset{a_{ij}}{\text{maximize}} \ \text{trace}\, S_b, \tag{4.2}$$

where $a_{ij}$ is the hard cluster assignment as $a_{ij} \in \{0,1\}, \Sigma_{j=1}^{k} a_{ij} = 1$ and

$$S_b = \sum_{j=1}^{k} n_j(\mu_j - \mu_0)(\mu_j - \mu_0)^T, \tag{4.3}$$

where $\mu_0$ is the global mean, $n_j = \Sigma_{i=1}^{N} a_{ij}$ is the number of samples in $\mathscr{C}_j$. Moreover, the within-cluster scatter $S_w$ and the total scatter $S_t$ are respectively given by

$$S_w = \sum_{j=1}^{k} \sum_{\mathbf{x}_i \in \mathscr{C}_j} (\mathbf{x}_i - \mu_j)(\mathbf{x}_i - \mu_j)^T, \tag{4.4}$$

$$S_t = \sum_{i=1}^{N} (\mathbf{x}_i - \mu_0)(\mathbf{x}_i - \mu_0)^T, \tag{4.5}$$

and

$$S_t = S_w + S_b \, . \tag{4.6}$$

Without loss of generality, we assume that the data $X \in \mathbb{R}^{M \times N}$ has been centered such that the global mean is $\mu_0 = 0$. To express $\mu_j$ in terms of $X$, we denote a scalar cluster membership matrix $A \in \mathbb{R}^{N \times K}$ as

$$A_{ij} = \begin{cases} \frac{1}{\sqrt{n_j}} & \text{if } \mathbf{x}_i \in C_j \\ 0 & \text{if } \mathbf{x}_i \notin C_j, \end{cases} \tag{4.7}$$

then $A^T A = I_k$ and the objective of $k$-means in (4.2) can be equivalently written as

$$\begin{aligned} &\underset{A}{\text{maximize}} \ \text{trace}\big(A^T X^T X A\big), \\ &\text{subject to } \ A^T A = I_k, \ \ A_{ij} \in \{0, \frac{1}{\sqrt{n_j}}\}. \end{aligned} \tag{4.8}$$

It is known that the discrete constraint in (4.8) makes the problem NP-hard to solve [10]. In literature, various methods have been proposed to the problem, such as the iterative descent method [12], the expectation-maximization method [3], the spectral relaxation method [7], and many others. In particular, the spectral relaxation method relaxes the discrete cluster memberships of $A$ to numerical values, thus (4.8) is relaxed to

$$\begin{aligned} &\underset{A}{\text{maximize}} \ \text{trace}\big(A^T X^T X A\big), \\ &\text{subject to } \ A^T A = I_k, \ \ A_{ij} \in \mathbb{R}. \end{aligned} \tag{4.9}$$

If $A$ reduces to a one column vector, the problem in (4.9) is exactly a Rayleigh quotient and the optimal $A^*$ is given by the eigenvector $\mathbf{u}_{max}$ in the largest eigenvalue pair $\{\lambda_{max}, \mathbf{u}_{max}\}$ of $X^T X$. If $A$ represents the relaxed assignment of multi-cluster memberships, according to the Ky Fan introduced in chapter 2, let the eigenvalues of $X^T X$ be ordered as $\lambda_{max} = \lambda_1 \geq, ..., \geq \lambda_N = \lambda_{min}$ and the corresponding eigenvectors as $\mathbf{u}_1, ..., \mathbf{u}_N$, then the optimal $A^*$ subject to $A^{*T} A^* = I_k$ is given by $A^* = U_k V$, where $U_k = [\mathbf{u}_1, ..., \mathbf{u}_k]$, and $V$ is an arbitrary $k \times k$ orthogonal matrix, and $\max \text{trace}\big(U^T X^T X U\big) = \lambda_1 + .. + \lambda_k$. Thus, for a given cluster number $k$, the $k$-means can be solved as an eigenvalue problem and the discrete cluster memberships of the original $A$ can be recovered using the iterative descend $k$-means method on $U_k$ or using the spectral ordering proposed by Ding and He [8].

## 4.3 Optimizing Multiple Kernels for *k*-means

We further generalize the objective in (4.9) by applying the feature map $\phi(\cdot): \mathbb{R} \rightarrow \mathscr{F}$ on $X$, then the centered data in Hilbert space $\mathscr{F}$ is denoted as $X^{\Phi}$, given by

$$X^{\Phi} = [\phi(\mathbf{x}_1) - \mu_0^{\Phi}, \phi(\mathbf{x}_2) - \mu_0^{\Phi}, ..., \phi(\mathbf{x}_N) - \mu_0^{\Phi}], \tag{4.10}$$

where $\phi(\mathbf{x}_i)$ is the feature map applied on the column vector of the $i$-th data point in $\mathscr{F}$, $\mu_0^{\Phi}$ is the global mean in $\mathscr{F}$. The inner product $X^TX$ corresponds to $X^{\Phi T}X^{\Phi}$ in Hilbert space and can be combined using the kernel trick $\mathscr{K}(\mathbf{x}_u, \mathbf{x}_v) = \phi(\mathbf{x}_u)^T\phi(\mathbf{x}_v)$, where $\mathscr{K}(\cdot,\cdot)$ is a Mercer kernel and $K$ is the corresponding kernel matrix. We denote $G$ as the centered kernel matrix as $G = PKP$, where $P$ is the centering matrix $P = I_N - (1/N)\mathbf{1}_N\mathbf{1}_N^T$, $I_N$ is the $N \times N$ identity matrix, $\mathbf{1}_N$ is a column vector of $N$ ones. Note that the trace of between-cluster scatter trace$(S_b^{\Phi})$ takes the form of a series of dot products in the centered Hilbert space, therefore the between-cluster scatter is equivalent to

$$\mathrm{trace}\left(S_b^{\Phi}\right) = \mathrm{trace}\left(A^TGA\right). \tag{4.11}$$

Now let's assume that $X_1, ..., X_p$ are $p$ different representations of the same $N$ objects. We are motivated to extend the clustering problem of a single data set to multiple data sets, thus we can easily combine the centered kernel matrices $G_r$, $(r = 1, ..., p)$ in a parametric linear additive manner as

$$\Omega = \left\{ \sum_{r=1}^{p} \theta_r G_r \;\middle|\; \forall \theta_r \geq 0, \; \sum_{r=1}^{p} \theta_r = 1 \right\}, \tag{4.12}$$

where $\theta_r$ are coefficients of the kernel matrices, $G_r$ are normalized kernel matrices [23] centered in the Hilbert space. Kernel normalization ensures that $\phi(\mathbf{x}_i)^T\phi(\mathbf{x}_i) = 1$ thus makes the kernels comparable to each other. The $k$-means objective in (4.9) is thus extended to $\mathscr{F}$ and multiple data sets are incorporated, given by

$$\boxed{\text{Q1:}} \quad \underset{A,\theta}{\text{maximize}} \; \mathrm{trace}\left(A^T\Omega A\right), \tag{4.13}$$

$$\text{subject to } \; A^TA = I_k, \; A_{ij} \in \mathbb{R},$$

$$\Omega = \sum_{r=1}^{p} \theta_r G_r,$$

$$\theta_r \geq 0, \; r = 1, ..., p$$

$$\sum_{r=1}^{p} \theta_r = 1.$$

The idea of combining multiple kernels in $k$-means clustering is also mentioned in some related work, for instance, the NAML algorithm proposed by Chen *et al.* [5].

In their approach, the objective of $k$-means is formulated as the trace of $S_w$ using the Mahalanobis distance. Following their mathematical derivations, the $k$-means objective on a single data source is given by

$$\begin{aligned} &\underset{A}{\text{maximize}} \ \text{trace}\left(A^T X^{\Phi T}\left(X^{\Phi} X^{\Phi T}+\lambda I_{inf}\right)^{-1} X^{\Phi} A\right), \\ &\text{subject to } \ A^T A = I_k, \ A_{ij} \in \mathbb{R}, \end{aligned} \tag{4.14}$$

where $A$ is also a relaxed cluster assignment matrix, $\lambda$ is a regularization parameter on the covariance matrix, $X^{\Phi}$ is the centered data defined identically as our paper. Unfortunately, the objective in (4.14) is not explicitly computable because $\phi(\mathbf{x})$ may be infinite dimensional, thus the covariance matrix $X^{\Phi}X^{\Phi T}$ is an infinite size matrix. To tackle this flaw, NAML algorithm involves linear discriminant analysis (LDA) to find a linear projection $J \in \mathbb{R}^{m \times d}$ that maps $\phi(\mathbf{x})$ from the infinite dimensional space in $\mathscr{F}$ to a $d$-dimensional space $\mathscr{D}$. According to the Representer Theorem [22], the optimal projection $J$ is in the span of the images of data points in $\mathscr{F}$ as $P = X^{\Phi}Q$, where $Q \in \mathbb{R}^{N \times d}$. Incorporating $J$ in the objective of $k$-means, (4.14) is extended as

$$\begin{aligned} &\underset{A,Q}{\text{maximize}} \ \text{trace}\left(A^T GQ\left(Q^T(GG+\lambda G)Q\right)^{-1} Q^T GA\right) \\ &\text{subject to } \ A^T A = I_k, \ A_{ij} \in \mathbb{R}, \end{aligned} \tag{4.15}$$

which now becomes explicitly computable because $X^{\Phi}$ is transformed as $G$ on the basis of kernel trick. However, to solve (4.15) one needs to optimize $A$ and $Q$ iteratively. To incorporate multiple centered kernels $G_1, ..., G_p$, in NAML, the objective in (4.15) is further extended as

$$\begin{aligned} &\underset{A,Q,\theta}{\text{maximize}} \ \text{trace}\left(A^T \Omega Q\left(Q^T(\Omega\Omega+\lambda \Omega)Q\right)^{-1} Q^T \Omega A\right) \\ &\text{subject to } \ A^T A = I_k, \ A_{ij} \in \mathbb{R}, \\ &\qquad \Omega = \sum_{r=1}^{p} \theta_r G_r, \\ &\qquad \theta_r \geq 0, \ \sum_{r=1}^{p} \theta_r = 1, \ r = 1, ..., p, \end{aligned} \tag{4.16}$$

which is a non-convex objective to optimize $A$, $Q$, and $\theta$ simultaneously. To solve this, Chen *et al.* optimize $A$, $Q$, and $\theta$ iteratively in a tri-level optimization procedure. Comparing with NAML, our proposed objective in (4.13) only optimizes $A$ and $\theta$, which is also a non-convex problem to solve $A$ and $\theta$ simultaneously. To solve this, we propose a simple alternative minimization technique [6] to solve $A$ and $\theta$ iteratively as a bi-level optimization procedure.

## 4.4 Bi-level Optimization of $k$-means on Multiple Kernels

To solve the objective in the problem Q1 (4.13), in the first phase we maximize $\mathscr{J}_{Q1}$ with respect to $A$, keeping $\theta$ fixed. In the second phase we maximize $\mathscr{J}_{Q1}$ with respect to $\theta$, keeping $A$ fixed. The two-stage optimization is then repeated until convergence (the proof will be shown later). When $\theta$ is fixed, the problem is exactly the relaxed $k$-means clustering problem as discussed before. When $A$ is fixed, the problem of maximizing $\mathscr{J}_{Q1}$ reduces to the optimization of the coefficients $\theta_r$ assigned on kernels with the given cluster memberships. We will show that when the memberships are given, the problem in Q1 can be formulated as KFD in $\mathscr{F}$.

### 4.4.1 *The Role of Cluster Assignment*

In problem Q1, when we maximize $\mathscr{J}_{Q1}$ with respect to $A$ using the fixed $\theta$, the obtained $N \times k$ weighted cluster indicator matrix $A$ can also be regarded as the one-vs-others (1vsA) coding of the cluster assignments because each column of $A$ actually distinguishes one cluster from the other clusters. When $A$ is given, the between-cluster scatter matrix $S_b^{\Phi}$ is fixed, thus the problem of optimizing the coefficients of multiple kernel matrices is equivalent to optimizing the KFD [21] using multiple kernel matrices. The scatter matrix of KFD is determined by the cluster assignments of the data points, which can be obtained via an affinity function using $A$ as the input, given by

$$F_{ij} = \begin{cases} +1 & \text{if } A_{ij} > 0, \;\; i = 1, ..., N, \; j = 1, ..., k \\ -1 & \text{if } A_{ij} = 0, \;\; i = 1, ..., N, \; j = 1, ..., k \end{cases}, \tag{4.17}$$

where $F$ is an affinity matrix using $\{+1, -1\}$ to discriminate the cluster assignments. In the second iteration of our algorithm, to maximize $\mathscr{J}_{Q1}$ with respect to $\theta$ using the fixed $A$, we formulate it as the optimization of KFD on multiple kernel matrices using the affinity matrix $F$ as input.

### 4.4.2 *Optimizing the Kernel Coefficients as KFD*

As known, for a single data set $X$, given labels of two classes, to find the linear discriminant in $\mathscr{F}$ we need to maximize

$$\underset{\mathbf{w}}{\text{maximize}} \; \frac{\mathbf{w}^T S_b^{\Phi} \mathbf{w}}{\mathbf{w}^T (S_w^{\Phi} + \rho I) \mathbf{w}}, \tag{4.18}$$

where $\mathbf{w}$ is the norm vector of the separating hyperplane in $\mathscr{F}$, $S_b^{\Phi}$ and $S_w^{\Phi}$ are respectively the between-class and the within-cluster scatters in $\mathscr{F}$, $\rho$ is the regularization term to ensure the positive definiteness of the denominator. For $k$ multiple classes, denote $W = [\mathbf{w}_1, ..., \mathbf{w}_k]$ as the matrix where each column corresponds to the discriminative direction of 1vsA classes. An important property to notice about the KFD objective is that it is invariant w.r.t. rescalings of $\mathbf{w}$ [9]. Hence, we can always choose $\mathbf{w}$ ($W$) such that the denominator is simply $\mathbf{w}^T S_w \mathbf{w} = 1$ ($W^T S_w W = I_k$) [12]. In other words, if the within-class scatter is isotropic, the norm vectors of discriminant projections are merely the eigenvectors of the between-class scatter [9]. Thus the objective of KFD can be simplified as a Rayleigh quotient formulation. Moreover, the vectors in $W$ are orthogonal to each other, therefore, for convenience we can further rescale $W$ to have $W^T W = I_k$ (it can be proved that rescaling $W$ does not change the clustering result). Thus, the KFD objective can be expressed as

$$\begin{aligned} &\underset{W}{\text{maximize}} \ \ \text{trace}\left(W^T S_b^{\Phi} W\right), \\ &\text{subject to} \ \ W^T W = I_k. \end{aligned} \tag{4.19}$$

Based on the observations above, we formulate the objective of $k$-means using multiple kernels as

$$\begin{aligned} \boxed{\text{Q2:}} \ &\underset{A,W,\theta}{\text{maximize}} \ \text{trace}\left(W^T A^T \Omega A W\right), \\ &\text{subject to} \ \ A^T A = I_k, \\ &\qquad\qquad\ W^T W = I_k, \\ &\qquad\qquad\ \Omega = \sum_{r=1}^{p} \theta_r G_r, \\ &\qquad\qquad\ \theta_r \geq 0, \ \ r = 1, ..., p \\ &\qquad\qquad\ \sum_{r=1}^{p} \theta_r = 1. \end{aligned} \tag{4.20}$$

In the $k$-means step, we set $W = I_k$ and optimize $A$ (it can be easily proved that fixing $W$ as $I_k$ does not change the clustering result). In the KFD step, we fix $A$ and optimize $W$ and $\theta$. Therefore, the two components optimize towards the same objective as a Rayleigh quotient in $\mathscr{F}$, which also guarantee that the iterative optimization of these two steps will necessary converge to a local optimum. Moreover, in the KFD step of the present problem, we are not interested in $W$ which determines the separating hyperplane, instead, we only need the optimal coefficients $\theta_r$ assigned on the multiple kernels. In particular, it is known that Fisher Discriminant Analysis is related to the least squares approach [9], and the KFD is related to the least squares support vector machines (LSSVM) proposed by Suykens *et al.* [26, 27]. Therefore we can solve the KFD problem as LSSVM using multiple kernels.

### 4.4.3 Solving KFD as LSSVM Using Multiple Kernels

In LSSVM, the cost function of the classification error is replaced by a least squares term [27] and the inequalities in the constraint are replaced by equalities, given by

$$\begin{aligned} \underset{\mathbf{w},b,\mathbf{e}}{\text{minimize}} \quad & \frac{1}{2}\mathbf{w}^T\mathbf{w} + \frac{1}{2}\lambda\mathbf{e}^T\mathbf{e} \\ \text{subject to} \quad & y_i[\mathbf{w}^T\phi(\mathbf{x}_i) + b] = 1 - e_i, \quad i = 1,...,N, \end{aligned} \tag{4.21}$$

where $\mathbf{w}$ is the norm vector of separating hyper-plane, $\mathbf{x}_i$ are data samples, $\phi(\cdot)$ is the feature map, $y_i$ are the cluster assignments represented in the affinity function $F$, $\lambda > 0$ is a positive regularization parameter, $\mathbf{e}$ are the least squares error terms. Taking the conditions for optimality from the Lagrangian, eliminating $\mathbf{w}, \mathbf{e}$, defining $\mathbf{y} = [y_1,...,y_N]^T$ and $Y = diag(y_1,...,y_N)$, one obtains the following linear system [26, 27]:

$$\left[\begin{array}{c|c} 0 & \mathbf{y}^T \\ \hline \mathbf{y} & YKY + I/\lambda \end{array}\right] \left[\begin{array}{c} b \\ \hline \alpha \end{array}\right] = \left[\begin{array}{c} 0 \\ \hline \mathbf{1} \end{array}\right], \tag{4.22}$$

where $\alpha$ are unconstrained dual variables, $K$ is the kernel matrix obtained by kernel trick as $\mathscr{K}(\mathbf{x}_i, \mathbf{x}_j) = \phi(\mathbf{x}_i)^T\phi(\mathbf{x}_j)$. Without loss of generality, we denote $\beta = Y\alpha$ such that (4.22) becomes

$$\left[\begin{array}{c|c} 0 & \mathbf{1}^T \\ \hline \mathbf{1} & K + Y^{-2}/\lambda \end{array}\right] \left[\begin{array}{c} b \\ \hline \beta \end{array}\right] = \left[\begin{array}{c} 0 \\ \hline Y^{-1}\mathbf{1} \end{array}\right]. \tag{4.23}$$

In (4.17), the elements of $F$ are $\{-1, +1\}$ such that we have $Y^{-2} = I_N$, the coefficient matrix of the linear system in (23) then becomes a static value in the multi-class case. In 1vsA coding, solving (4.22) requires to solve $k$ number of linear problems whereas in (4.22), the coefficient matrix is only factorized once such that the solution of $\beta_q$ w.r.t. the multi-class assignments $\mathbf{y}_q$ is very efficient to obtain. To incorporate multiple kernels, we follow the multiple kernel learning (MKL) approach proposed by Lanckriet *et al.* [16] and formulate the LSSVM MKL as a QCQP problem, given by

$$\begin{aligned} \underset{\beta,t}{\text{minimize}} \quad & \frac{1}{2}t + \frac{1}{2\lambda}\beta^T\beta - \beta^T Y^{-1}\mathbf{1} \\ \text{subject to} \quad & \sum_{i=1}^{N} \beta_i = 0, \\ & t \geq \beta^T K_r \beta, \quad r = 1,...,p. \end{aligned} \tag{4.24}$$

In our problem, we use the normalized and centered kernel matrices of all samples thus $K_r$ is equal to $G_r$, the kernel coefficients $\theta_r$ correspond to the dual variables bounded by the quadratic constraints in (4.24). The column vector of $F$, denoted as $F_j$, $j = 1,...,k$ correspond to the $k$ number of $Y_1,...,Y_k$ in (4.23), where $Y_j = diag(F_j)$, $j = 1,...,k$. The bias term $b$ can be solved independently using the optimal

$\beta^*$ and the optimal $\theta^*$, thus can be dropped out from (4.24). Concluding the previous discussion, the SIP formulation of the LSSVM MKL is given by

$$\begin{aligned}
&\max_{\theta,u} u \\
&\text{s.t. } \theta_r \geq 0,\ r = 1,...,p+1 \\
&\quad \sum_{r=1}^{p+1} \theta_r = 1, \\
&\quad \sum_{r=1}^{p+1} \theta_r f_r(\beta) \geq u,\ \forall \beta \\
&\quad f_r(\beta) = \sum_{q=1}^{k} \left( \frac{1}{2} \beta_q^T G_r \beta_q - \beta_q^T Y_q^{-1} \mathbf{1} \right),\ r = 1,...,p+1.
\end{aligned} \tag{4.25}$$

The pseudocode to solve the LSSVM MKL in (4.25) is presented as follows:

---

**Algorithm 4.4.1.** SIP-LS-SVM-MKL($G_1,...,G_p,F$)

---

Obtain the initial guess $\beta^{(0)} = [\beta_1^{(0)},...,\beta_k^{(0)}]$
$\tau = 0$
**while** $(\Delta u > \varepsilon)$
**do**
$step1$ : Fix $\beta$, solve $\theta^{(\tau)}$then obtain $u^{(\tau)}$
$step2$ : Compute the kernel combination $\Omega^{(\tau)}$
$step3$ : Solve the single LS-SVM for the optimal $\beta^{(\tau)}$
$step4$ : Compute $f_1(\beta^{(\tau)}),...,f_{p+1}(\beta^{(\tau)})$
$step5$ : $\Delta u = |1 - \frac{\sum_{j=1}^{p+1} \theta_j^{(\tau)} f_j(\beta^{(\tau)})}{u^{(\tau)}}|$
$step6$ : $\tau := \tau + 1$
**comment:** $\tau$ is the indicator of the current loop
**return** $(\theta^{(\tau)}, \beta^{(\tau)})$

---

In Algorithm 4.4.1 $G_1,...,G_p$ are centered kernel matrices of multiple sources, an identity matrix is set as $G_{p+1}$ to estimate the regularization parameter, $Y_1,...,Y_k$ are the $N \times N$ diagonal matrices constructed from $F$. The $\varepsilon$ is a fixed constant as the stopping rule of SIP iterations and is set empirically as 0.0001 in our implementation. Normally the SIP takes about ten iterations to converge. In Algorithm 4.1, Step 1 optimizes $\theta$ as a linear programming and Step 3 is simply a linear problem as

$$\left[ \begin{array}{c|c} 0 & \mathbf{1}^T \\ \hline \mathbf{1} & \Omega^{(\tau)} \end{array} \right] \left[ \begin{array}{c} b^{(\tau)} \\ \hline \beta^{(\tau)} \end{array} \right] = \left[ \begin{array}{c} 0 \\ \hline Y^{-1}\mathbf{1} \end{array} \right], \tag{4.26}$$

where $\Omega^{(\tau)} = \sum_{r=1}^{p+1} \theta_j^{(\tau)} G_r$.

### 4.4.4 Optimized Data Fusion for Kernel k-means Clustering (OKKC)

Now we have clarified the two algorithmic components to optimize the objective Q2 as defined in (4.20). The main characteristic is that the cluster assignments and the coefficients of kernels are optimized iteratively and adaptively until convergence. The coefficients assigned on multiple kernel matrices leverage the effect of different kernels in data integration to optimize the objective of clustering. Comparing to the average combination of kernel matrices, the optimized combination approach may be more robust to noisy and irrelevant data sources. We name the proposed algorithm optimized kernel *k*-means clustering (OKKC) and its pseudocode is presented in Algorithm 4.4.2 as follows:

---

**Algorithm 4.4.2.** OKKC$(G_1, G_2, ..., G_p, k)$

---

**comment:** Obtain the $\Omega^{(0)}$ by the initial guess of $\theta_1^{(0)}, ..., \theta_p^{(0)}$

$A^{(0)} \leftarrow$ KERNEL K-MEANS $(\Omega^{(0)}, k)$

$\gamma = 0$

**while** $(\Delta A > \varepsilon)$

$$\textbf{do} \begin{cases} step1 : F^{(\gamma)} \leftarrow A^{(\gamma)} \\ step2 : \Omega^{(\gamma+1)} \leftarrow \text{SIP-LS-SVM-MKL}(G_1, G_2, ..., G_p, F^{(\gamma)}) \\ step3 : A^{(\gamma+1)} \leftarrow \text{KERNEL K-MEANS}(\Omega^{(\gamma+1)}, k) \\ step4 : \Delta A = ||A^{(\gamma+1)} - A^{(\gamma)}||^2 / ||A^{(\gamma+1)}||^2 \\ step5 : \gamma := \gamma + 1 \end{cases}$$

**return** $(A^{(\gamma)}, \theta_1^{(\gamma)}, ..., \theta_p^{(\gamma)})$

---

The iteration in Algorithm 4.4.2 terminates when the relaxed cluster membership matrix $A$ stops changing. The tolerance value $\varepsilon$ is constant value as the stopping rule of OKKC and in our implementation it is set to 0.05. In our implementation, the final cluster assignment is obtained using the kernel *k*-means algorithm [11] on the optimally combined kernel matrix $\Omega^{(\tau)}$.

### 4.4.5 Computational Complexity

The proposed OKKC algorithm has several advantages over some similar algorithms proposed in the literature. The optimization procedure of OKKC is bi-level, which is simpler than the tri-level architecture of the NAML algorithm. The kernel coefficients in OKKC is optimized as LS-SVM MKL, which can be solved efficiently as a convex SIP problem. The kernel coefficients are obtained as iterations of two linear systems: a single kernel LSSVM problem and a linear problem to optimize the kernel coefficients. The time complexity of OKKC is $O\{\gamma[N^3 + \tau(N^2 + p^3)] + lkN^2\}$, where $\gamma$ is the number of OKKC iterations, $O(N^3)$

is the complexity of eigenvalue decomposition, $\tau$ is the number of SIP iterations, the complexity of LS-SVM based on conjugate gradient method is $O(N^2)$, the complexity of optimizing kernel coefficients is $O(p^3)$, $l$ is the fixed iteration of $k$-means clustering, $p$ is the number of kernels, and $O(lkN^2)$ is the complexity of $k$-means to finally obtain the cluster assignment. In contrast, the complexity of NAML algorithm is $O\{\gamma(N^3 + N^3 + pk^2N^2 + pk^3N^3)\}$, where the complexities of obtaining cluster assignment and projection are all $O(N^3)$, the complexity of solving QCQP based problem is $O(pk^2N^2 + pk^3N^3)$, and $k$ is the number of clusters. Obviously, the complexity of OKKC is much smaller than NAML.

## 4.5 Experimental Results

The proposed algorithm is evaluated on public data sets and real application data to study the empirical performance. In particular, we systematically compare it with the NAML algorithm on clustering performance, computational efficiency and the effect of data fusion.

**Table 4.1** Summary of the data sets

| Data set | Dimension | Instance | Class | Kernel function | Nr. of kernels |
|---|---|---|---|---|---|
| iris | 4 | 150 | 3 | RBF | 10 |
| wine | 13 | 178 | 3 | RBF | 10 |
| yeast | 17 | 384 | 5 | RBF | 10 |
| satimage | 36 | 480 | 6 | RBF | 10 |
| pen digit | 16 | 800 | 10 | RBF | 10 |
| disease | - | 620 | 2 | - | 9 |
| GO | 7403 | 620 | 2 | linear | |
| MeSH | 15569 | 620 | 2 | linear | |
| OMIM | 3402 | 620 | 2 | linear | |
| LDDB | 890 | 620 | 2 | linear | |
| eVOC | 1659 | 620 | 2 | linear | |
| KO | 554 | 620 | 2 | linear | |
| MPO | 3446 | 620 | 2 | linear | |
| Uniprot | 520 | 620 | 2 | linear | |
| journal | 669860 | 1424 | 7 | linear | 4 |

### 4.5.1 *Data Sets and Experimental Settings*

We adopt five data sets from the UCI machine learning repository and two data sets from real-life bioinformatics and scientometrics applications. The five UCI data sets are: Iris, Wine, Yeast, Satimage and Pen digit recognition. The original Satimage and Pen digit data contain a large amount of data points, so we sample 80 data points from each class and construct the data sets. For each data set, we generate ten

RBF kernel matrices using different kernel widths $\sigma$ in the RBF function $\kappa(\mathbf{x}_i,\mathbf{x}_j) = \exp(-||\mathbf{x}_i - \mathbf{x}_j||^2/2\sigma^2)$. We denote the average sample covariance of data set as $c$, then the $\sigma$ values of the RBF kernels are respectively equal to $\{\frac{1}{4}c, \frac{1}{2}c, c, ..., 7c, 8c\}$. These ten kernel matrices are combined to simulate a kernel fusion problem for clustering analysis.

We also apply the proposed algorithm on data sets of two real applications. The first data set is taken from a bioinformatics application using biomedical text mining to cluster disease relevant genes [33]. The details will be presented in the next chapter. We select controlled vocabularies (CVs) from nine bio-ontologies for text mining and store the terms as bag-of-words respectively. The nine CVocs are used to index the title and abstract of around 290,000 human gene-related publications in MEDLINE to construct the doc-by-term vectors. According to the mapping of genes and publications in Entrez GeneRIF, the doc-by-term vectors are averagely combined as gene-by-term vectors, which are denoted as the term profiles of genes and proteins. The term profiles are distinguished by the bio-ontologies where the CVocs are selected and labeled as GO, MeSH, OMIM, LDDB, eVOC, KO, MPO, SNOMED and UniProtKB. Using these term profiles, we evaluate the performance of clustering a benchmark data set consisting of 620 disease relevant genes categorized in 29 genetic diseases. The numbers of genes categorized in the diseases are very imbalanced, moreover, some genes are simultaneously related to several diseases. To obtain meaningful clusters and evaluations, we enumerate all the pairwise combinations of the 29 diseases (406 combinations). In each run, the related genes of each paired diseases combination are selected and clustered into two groups, then the performance is evaluated using the disease labels. The genes related to both diseases in the paired combination are removed before clustering (totally there are less than 5% genes being removed). Finally, the average performance of all the 406 paired combinations is used as the overall clustering performance.

The second real-life data set is taken from a scientometrics application [18]. The raw experimental data contains more than six million published papers from 2002 to 2006 (e.g., articles, letters, notes, reviews) indexed in the Web of Science (WoS) data based provided by Thomson Scientific. In our preliminary study of clustering of journal sets, the titles, abstracts and keywords of the journal publications are indexed by text mining program using no controlled vocabulary. The index contains 9,473,601 terms and we cut the Zipf curve [34] of the indexed terms at the head and the tail to remove the rare terms, stopwords and common words, which are known as usually irrelevant, also noisy for the clustering purpose. After the Zipf cut, 669,860 terms are used to represent the journal publications in vector space models where the terms are attributes and the weights are calculated by four weighting schemes: TF-IDF, IDF, TF and binary. The publication-by-term vectors are then aggregated to journal-by-term vectors as the representations of journal data. From the WoS database, we refer to the Essential Science Index (ESI) labels and select 1424 journals as the experimental data in this paper. The distributions of ESI labels of these journals are balanced because we want to avoid the affect of skewed distributions in cluster evaluation. In experiment, we cluster the 1424 journals simultaneously into 7 clusters and evaluate the results with the ESI labels.

We summarize the number of samples, classes, dimensions and the number of combined kernels in Table 4.1. The disease and journal data sets have very high dimensionality so the kernel matrices are constructed using the linear kernel functions. The kernel matrices are normalized and centered. The data sets used in experiments are provided with labels, therefore the performance is evaluated as comparing the automatic partitions with the labels using Adjusted Rand Index (ARI) [14] and Normalized Mutual Information (NMI) [25].

### *4.5.2 Results*

The overall clustering results are shown in Table 4.2. For each data set, we present the best and the worst performance of clustering obtained on single kernel matrix. We compared three different approaches to combine multiple kernel matrices: the average combination of all kernel matrices in kernel $k$-means clustering, the proposed OKKC algorithm and the NAML algorithm. As shown, the performance obtained by OKKC is comparable to the results of the best individual kernel matrices. OKKC is also comparable to NAML on all the data sets, moreover, on Wine, Pen, Disease, and Journal data, OKKC performs significantly better than NAML (as shown in Table 4.3). The computational time used by OKKC is also smaller than NAML. Since OKKC and NAML use almost the same number of iterations to converge, the efficiency of OKKC is mainly brought by its bi-level optimization procedure and the linear system solution based on SIP formulation. In contrast, NAML optimizes three variables in a tri-level procedure and involves many inverse computation and eigenvalue decompositions on kernel matrices. Furthermore, in NAML, the kernel coefficients are optimized as a QCQP problem. When the number of data points and the number of classes are large, QCQP problem may have memory issues. In our experiment, when clustering Pen digit data and Journal data, the QCQP problem causes memory overflow on a laptop computer. Thus we had to solve them on a Unix system with larger amount of memory. On the contrary, the SIP formulation used in OKKC significantly reduces the computational burden of optimization and the clustering problem usually takes 25 to 35 minutes on an ordinary laptop.

We also compared the kernel coefficients optimized by OKKC and NAML on all the data sets. As shown in Figure 4.1, NAML algorithm often selects a single kernel for clustering (a sparse solution for data fusion). In contrast, OKKC algorithm often combines two or three kernel matrices in clustering. When combining $p$ kernel matrices, the regularization parameters $\lambda$ estimated in OKKC are shown as the coefficients of an additional $(p+1)$-th identity matrix (the last bar in the figures, except on disease data because $\lambda$ is also pre-selected), moreover, in OKKC it is easy to see that $\lambda = (\sum_{r=1}^{p} \theta_r)/\theta_{p+1}$. The $\lambda$ values of NAML are selected empirically according to the clustering performance. Practically, to determine the optimal regularization parameter in clustering analysis is hard because the data is unlabeled thus the model cannot be validated. Therefore, the automatic estimation of $\lambda$ in OKKC is useful and reliable in clustering.

| | best individual | | worst individual | | average combine | | | OKKC | | | | NAML | | | |
|---|---|---|---|---|---|---|---|---|---|---|---|---|---|---|---|
| | ARI | NMI | ARI | NMI | ARI | NMI | time(sec) | ARI | NMI | itr | time(sec) | ARI | NMI | itr | time(sec) |
| Iris | 0.7302<br>(0.0690) | 0.7637<br>(0.0606) | 0.6412<br>(0.1007) | 0.7047<br>(0.0543) | 0.7132<br>(0.1031) | 0.7641<br>(0.0414) | 0.22<br>(0.13) | 0.7516<br>(0.0690) | 0.7637<br>(0.0606) | 7.8<br>(3.7) | 5.32<br>(2.46) | 0.7464<br>(0.0207) | 0.7709<br>(0.0117) | 9.2<br>(2.5) | 15.45<br>(6.58) |
| Wine | 0.3489<br>(0.0887) | 0.3567<br>(0.0808) | 0.0387<br>(0.0175) | 0.0522<br>(0.0193) | 0.3188<br>(0.1264) | 0.3343<br>(0.1078) | 0.25<br>(0.03) | 0.3782<br>(0.0547) | 0.3955<br>(0.0527) | 10<br>(4.0) | 18.41<br>(11.35) | 0.2861<br>(0.1357) | 0.3053<br>(0.1206) | 6.7<br>(1.4) | 16.92<br>(3.87) |
| Yeast | 0.4246<br>(0.0554) | 0.5022<br>(0.0222) | 0.0007<br>(0.0025) | 0.0127<br>(0.0038) | 0.4193<br>(0.0529) | 0.4994<br>(0.0271) | 2.47<br>(0.05) | 0.4049<br>(0.0375) | 0.4867<br>(0.0193) | 7<br>(1.7) | 81.85<br>(14.58) | 0.4256<br>(0.0503) | 0.4998<br>(0.0167) | 10<br>(2) | 158.20<br>(30.38) |
| Satimage | 0.4765<br>(0.0515) | 0.5922<br>(0.0383) | 0.0004<br>(0.0024) | 0.0142<br>(0.0033) | 0.4891<br>(0.0476) | 0.6009<br>(0.0278) | 4.54<br>(0.07) | 0.4996<br>(0.0571) | 0.6004<br>(0.0415) | 10.2<br>(3.6) | 213.40<br>(98.70) | 0.4911<br>(0.0522) | 0.6027<br>(0.0307) | 8<br>(0.7) | 302<br>(55.65) |
| Pen digit | 0.5818<br>(0.0381) | 0.7169<br>(0.0174) | 0.2456<br>(0.0274) | 0.5659<br>(0.0257) | 0.5880<br>(0.0531) | 0.7201<br>(0.0295) | 15.95<br>(0.08) | 0.5904<br>(0.0459) | 0.7461<br>(0.0267) | 8<br>(4.38) | 396.48<br>(237.51) | 0.5723<br>(0.0492) | 0.7165<br>(0.0295) | 8<br>(4.2) | 1360.32<br>(583.74) |
| Disease genes | 0.7585<br>(0.0043) | 0.5281<br>(0.0078) | 0.5900<br>(0.0014) | 0.1928<br>(0.0042) | 0.7306<br>(0.0061) | 0.4702<br>(0.0101) | 931.98<br>(1.51) | 0.7641<br>(0.0078) | 0.5395<br>(0.0147) | 5<br>(1.5) | 1278.58<br>(120.35) | 0.7310<br>(0.0049) | 0.4715<br>(0.0089) | 8.5<br>(2.6) | 3268.83<br>(541.92) |
| Journal sets | 0.6644<br>(0.0878) | 0.7203<br>(0.0523) | 0.5341<br>0.0580 | 0.6472<br>0.0369 | 0.6774<br>(0.0316) | 0.7458<br>(0.0268) | 63.29<br>(1.21) | 0.6812<br>(0.0602) | 0.7420<br>(0.0439) | 8.2<br>(4.4) | 1829.39<br>(772.52) | 0.6294<br>(0.0535) | 0.7108<br>(0.0355) | 9.1<br>(6.1) | 4935.23<br>(3619.50) |

All the results are mean values of 20 random repetitions and the standard deviation (in parentheses). The individual kernels and average kernels are clustered using kernel $k$-means [11]. The OKKC is programmed using kernel $k$-means [11], Matlab functions *linsolve* and *linprog*. The disease gene data is clustered by OKKC using the explicit regularization parameter $\lambda$ (set to 0.0078) because the linear kernel matrices constructed from gene-by-term profiles are very sparse (a gene normally is only indexed by a small number of terms in the high dimensional vector space). In this case, the joint estimation assigns dominant coefficients on the identity matrix and decreases the clustering performance. The optimal $\lambda$ value is selected among ten values uniformly distributed on the log scale from $2^{-5}$ to $2^{-4}$. For other data sets, the $\lambda$ values are estimated automatically and their values are shown as $\lambda_{okkc}$ in Figure 1. The NAML is programmed as the algorithm proposed in [5] using Matlab and MOSEK [1]. We try forty-one different $\lambda$ values for NAML on the log scale from $2^{-20}$ to $2^{20}$ and the highest mean values and their deviations are presented. In general, the performance of NAML is not very sensitive to the $\lambda$ values. The optimal $\lambda$ values for NAML are shown in Figure 4.1 as $\lambda_{naml}$. The computational time (no underline) is evaluated on Matlab v7.6.0 + Windows XP SP2 installed on a Laptop computer with Intel Core 2 Duo 2.26GHz CPU and 2G memory. The computational time (underlined) is evaluated on Matlab v7.9.0 installed on a dual Opteron 250 Unix system with 7Gb memory.

**Table 4.2** Significance test of clustering performance

| data | OKKC vs. single | | OKKC vs. NAML | | OKKC vs. average | |
|---|---|---|---|---|---|---|
| | ARI | NMI | ARI | NMI | ARI | NMI |
| iris | 0.2213 | 0.8828 | 0.7131 | 0.5754 | 0.2282 | 0.9825 |
| wine | 0.2616 | 0.1029 | **0.0085**(+) | **0.0048**(+) | 0.0507 | **0.0262**(+) |
| yeast | 0.1648 | **0.0325**(-) | 0.1085 | **0.0342**(-) | 0.2913 | **0.0186**(-) |
| satimage | 0.1780 | 0.4845 | 0.6075 | 0.8284 | 0.5555 | 0.9635 |
| pen | **0.0154**(+) | 0.2534 | **3.9e-11**(+) | **3.7e-04**(+) | 0.4277 | **0.0035**(+) |
| disease | **1.3e-05**(+) | **1.9e-05**(+) | **4.6e-11**(+) | **3.0e-13**(+) | **7.8e-11**(+) | **1.6e-12**(+) |
| journal | 0.4963 | 0.2107 | **0.0114**(+) | **0.0096**(+) | 0.8375 | 0.7626 |

The presented numbers are $p$ values evaluated by paired t-tests on 20 random repetitions. The experimental settings are mentioned in Table II. The null hypothesis (the average performance of the comparing approaches is the same) is rejected at the significance level of 0.05. "+" represents that the performance of OKKC is higher than the comparing approaches. "-" means that the performance of OKKC is lower.

When using spectral relaxation, the optimal cluster number of $k$-means can be estimated by checking the plot of eigenvalues [30]. We can also use the same technique to find the optimal cluster number of data fusion using OKKC. To demonstrate this, we clustered all the data sets using different $k$ values and plot the eigenvalues in Figure 4.2. As shown, the obtained eigenvalues with various $k$ are slightly different with each other because a different $k$ yields a different set of optimized kernel coefficients. However, we also find that even the kernel fusion results are different, the plots of eigenvalues obtained from the combined kernel matrix are quite similar to each other. In practical explorative analysis, one may be able to determine the optimal and consistent cluster number using OKKC with various $k$ values. The results show that OKKC can also be applied to find the clusters using the eigenvalues.

## 4.6 Summary

This chapter presented OKKC, a data fusion algorithm for kernel $k$-means clustering, where the coefficients of kernel matrices in the combination are optimized automatically. The proposed algorithm extends the classical $k$-means clustering algorithm in Hilbert space, where multiple heterogeneous data sats are represented as kernel matrices and combined for data fusion. The objective of OKKC is formulated as a Rayleigh quotient function of two variables, the cluster assignment $A$ and the kernel coefficients $\theta$, which are optimized iteratively towards the same objective. The proposed algorithm is shown to converge locally and implemented as an integration of kernel $k$-means clustering and LSSVM MKL.

The experimental results on UCI data sets and real application data sets validated the proposed method. The proposed OKKC algorithm obtained comparable result with the best individual kernel matrix and the NAML algorithm. Moreover, in

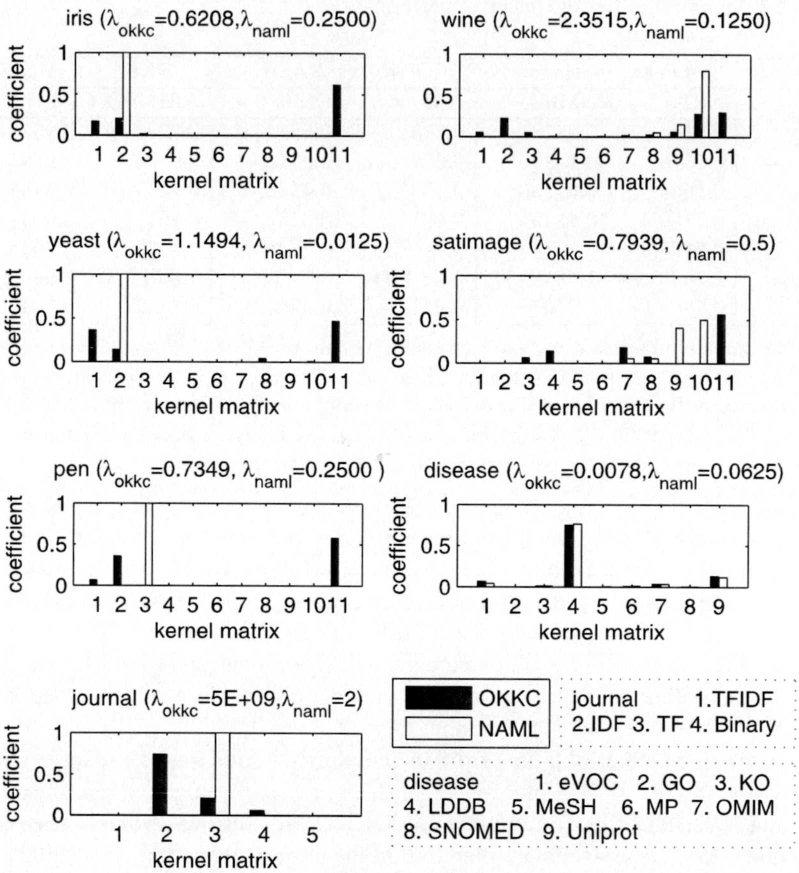


**Fig. 4.1** Kernel coefficients learned by OKKC and NAML. For OKKC applied on iris, wine, yeast, satimage, pen and journal data, the last coefficients correspond to the inverse values of the regularization parameters

several data sets it performs significantly better. Because of its simple optimization procedure and low computational complexity, the computational time of OKKC is always smaller than the NAML. The proposed algorithm also scales up well on large data sets thus it is more easy to run on ordinary machines.

In future work, the proposed algorithm is also possible to incorporate OKKC with overlapping cluster membership, known as "soft clustering". In many applications such as bioinformatics, a gene or protein may be simultaneously related to several biomedical concepts so it is necessary to have a "soft clustering" algorithm to combine multiple data sources.

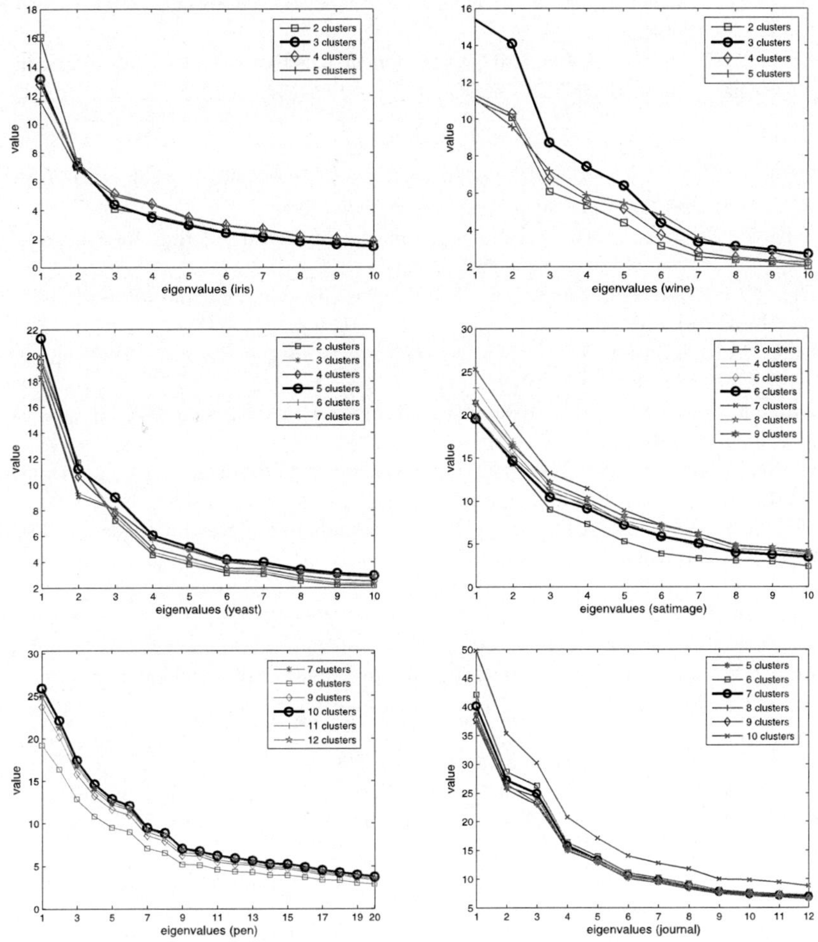


**Fig. 4.2** Eigenvalues of optimally combined kernels of data sets obtained by OKKC. For each data set we try four to six $k$ values including the one suggested by the reference labels, which is shown as a bold dark line, other values are shown as grey lines. The eigenvalues in disease gene clustering are not shown because there are 406 different clustering tasks.

# References

1. Andersen, E.D., Andersen, K.D.: The MOSEK interior point optimizer for linear programming: an implementation of the homogeneous algorithm. In: High Perf. Optimization, pp. 197–232. Kluwer Academic Publishers, New York (2000)
2. Bhatia, R.: Matrix Analysis. Springer, New York (1997)
3. Bishop, C.M.: Pattern recognition and machine learning. Springer, New York (2006)

4. Boyd, S., Vandenberghe, L.: Convex Optimization. Cambridge University Press, Cambridge (2004)
5. Chen, J., Zhao, Z., Ye, J.P., Liu, H.: Nonlinear adaptive distance metric learning for clustering. In: Proc. of ACM SIGKDD 2007, pp. 123–132. ACM Press, New York (2007)
6. Csiszar, I., Tusnady, G.: Information geometry and alternating minimization procedures. Statistics and Decisions suppl. 1, 205–237 (1984)
7. Ding, C., He, X.: K-means Clustering via Principal Component Analysis. In: Proc. of ICML 2004, pp. 225–232. ACM Press, New York (2004)
8. Ding, C., He, X.: Linearized cluster assignment via spectral ordering. In: Proc. of ICML 2004, vol. 30. ACM Press, New York (2004)
9. Duda, R.O., Hart, P.E., Stork, D.G.: Pattern Classification, 2nd edn. John Wiley & Sons Inc., New York (2001)
10. Garey, M.R., Johnson, D.S.: Computers and Intractability: A Guide to NP-Completeness. W. H. Freeman, New York (1979)
11. Girolami, M.: Mercer Kernel-Based Clustering in Feature Space. IEEE Trans. Neural Networks 13, 780–784 (2002)
12. Hastie, T., Tibshirani, R., Friedman, J.: The Elements of Statistical Learning: Data Mining, Inference, and Prediction, 2nd edn. Springer, Heidelberg (2009)
13. Hettich, R., Kortanek, K.O.: Semi-infinite programming: theory, methods, and applications. SIAM Review 35, 380–429 (1993)
14. Hubert, L., Arabie, P.: Comparing partitions. Journal of Classification 2, 193–218 (1985)
15. Jain, A.K., Dubes, R.C.: Algorithms for clustering data. Prentice Hall, New Jersey (1988)
16. Lanckriet, G.R.G., Cristianini, N., Bartlett, P., Ghaoui, L.E., Jordan, M.I.: Learning the Kernel Matrix with Semidefinite Programming. Journal of Machine Learning Reserach 5, 27–72 (2005)
17. Lange, T., Buhmann, J.M.: Fusion of Similarity Data in Clustering. Advances in Neural Information Processing Systems 18, 723–730 (2006)
18. Liu, X., Yu, S., Moreau, Y., De Moor, B., Glänzel, W., Janssens, F.: Hybrid Clustering of Text Mining and Bibliometrics Applied to Journal Sets. In: Proc. of the SIAM Data Mining Conference 2009. SIAM Press, Philadelphia (2009)
19. Ma, J., Sancho-Gómez, J.L., Ahalt, S.C.: Nonlinear Multiclass Discriminant Analysis. IEEE Signal Processing Letters 10, 196–199 (2003)
20. MacKay, D.J.C.: Information Theory, Inference, and Learning Algorithms. Cambridge University, Cambridge (2003)
21. Mika, S., Rätsch, G., Weston, J., Schölkopf, B.: Fisher discriminant analysis with kernels. In: IEEE Neural Networks for Signal Processing IX: Proceedings of the 1999 IEEE Signal Processing Society Workshop, pp. 41–48 (1999)
22. Schölkopf, B., Herbrich, R., Smola, A.J.: A generalized representer theorem. In: Helmbold, D.P., Williamson, B. (eds.) COLT 2001 and EuroCOLT 2001. LNCS (LNAI), vol. 2111, pp. 416–426. Springer, Heidelberg (2001)
23. Shawe-Taylor, J., Cristianin, N.: Kernel Methods for Pattern Analysis. Cambridge University Press, Cambridge (2004)
24. Stewart, G.W., Sun, J.G.: Matrix perturbation theory. Academic Press, Boston (1999)
25. Strehl, A., Ghosh, J.: Cluster Ensembles: A Knowledge Reuse Framework for Combining Multiple Partitions. Journal of Machine Learning Research 3, 583–617 (2002)
26. Suykens, J.A.K., Vandewalle, J.: Least squares support vector machine classifiers. Neural Processing Letters 9, 293–300 (1999)
27. Suykens, J.A.K., Van Gestel, T., De Brabanter, J., De Moor, B., Vandewalle, J.: Least Squares Support Vector Machines. World Scientific Press, Singapore (2002)

28. Sonnenburg, S., Rätsch, G., Schäfer, C., Schölkopf, B.: Large Scale Multiple Kernel Learning. Jounral of Machine Learning Research 7, 1531–1565 (2006)
29. Theodoridis, S., Koutroumbas, K.: Pattern Recognition, 2nd edn. Academic Press, London (2003)
30. von Luxburg, U.: A tutorial on spectral clustering. Statistics and Computing 17, 395–416 (2007)
31. Ye, J.P., Ji, S.W., Chen, J.H.: Multi-class Discriminant Kernel Learning via Convex Programming. Jounral of Machine Learning Research 9, 719–758 (2008)
32. Ye, J.P., Zhao, Z., Wu, M.: Discriminative K-Means for Clustering. Advances in Neural Information Processing Systems 19, 1649–1656 (2007)
33. Yu, S., Tranchevent, L., De Moor, B., Moreau, Y.: Gene prioritization and clustering by multi-view text mining. BMC Bioinformatics 11, 1–48 (2010)
34. Zipf, G.K.: Human behaviour and the principle of least effort: An introduction to human ecology. Hafner reprint, New York (1949)

# Chapter 5
# Multi-view Text Mining for Disease Gene Prioritization and Clustering

*So oft in theologic wars,*
*The disputants, I ween,*
*Rail on in utter ignorance*
*Of what each other mean,*
*And prate about an Elephant*
*Not one of them has seen!*
– "The Blind Men and the Elephant",
*John Godfrey Saxe (1816-1887)* –

## 5.1 Introduction

Text mining helps biologists to collect disease-gene associations automatically from large volumes of biological literature. During the past ten years, there was a surge of interests in automatic exploration of the biomedical literature, ranging from the modest approach of annotating and extracting keywords from text to more ambitious attempts such as Natural Language Processing, text-mining based network construction and inference. In particular, these efforts effectively help biologists to identify the most likely disease candidates for further experimental validation. The most important resource for text mining applications now is the MEDLINE database developed by the National Center for Biotechnology Information (NCBI) at the National Library of Medicine (NLM). MEDLINE covers all aspects of biology, chemistry, and medicine, there is almost no limit to the types of information that may be recovered through careful and exhaustive mining [45]. Therefore, a successful text mining approach relies much on an appropriate model. To create a text mining model, the selection of *Controlled Vocabulary* (CV) and the representation schemes of terms occupy a central role and the efficiency of biomedical knowledge discovery varies greatly between different text mining models. To address these challenges, we propose a *multi-view text mining* approach to retrieve information from different biomedical domain levels and combine them to identify the disease relevant genes

S. Yu et al.: Kernel-based Data Fusion for Machine Learning, SCI 345, pp. 109–144.
springerlink.com 

through prioritization. The *view* represents a text mining result retrieved by a specific CV, so the concept of *multi-view text mining* is featured as applying multiple controlled vocabularies to retrieve the gene-centric perspectives from free text publications. Since all the information is retrieved from the same MEDLINE database but only varied by the CV, the term *view* also indicates that the data consists of multiple domain-based perspectives of the same corpus. We expect that the correlated and complementary information contained in the *multi-view* textual data can facilitate the understanding about the roles of genes in genetic diseases.

## 5.2 Background: Computational Gene Prioritization

Genome-wide experimental methods to identify disease causing genes, such as linkage analysis and association studies, are often overwhelmed by large sets of candidate genes produced by the high throughput techniques [43]. In contrast, the low-throughput validation of candidate disease genes is time-consuming and expensive. Computational prioritization methods can rank candidate disease genes from these gene sets according their likeliness of being involved in a certain disease. Moreover, a systematic gene prioritization approach that integrates multiple genomic data sets provides a comprehensive *in silico* analysis on the basis of multiple knowledge sources. Several computational gene prioritization applications have been previously described. GeneSeeker [17] provides a web interface that filters candidate disease genes on the basis of cytogenetic location, phenotypes, and expression patterns. DGP (Disease Gene Prediction) [33] assigns probabilities to genes based on sequence properties that indicate their likelihood to the patterns of pathogenic mutations of certain monogenetic hereditary disease. PROSPECTR [2] also classifies disease genes by sequence information but uses a decision tree model. SUSPECTS [1] integrates the results of PROSPECTR with annotation data from Gene Ontology (GO), InterPro, and expression libraries to rank genes according to the likelihood that they are involved in a particular disorder. G2D (Candidate Genes to Inherited Diseases) [40] scores all concepts in GO according to their relevance to each disease via text mining. Then, candidate genes are scored through a BLASTX search on reference sequence. POCUS [56] exploits the tendency for genes to be involved in the same disease by identifiable similarities, such as shared GO annotation, shared InterPro domains or a similar expression profile. eVOC annotation [53] is a text mining approach that performs candidate gene selection using the eVOC ontology as a controlled vocabulary. It first associates eVOC terms and disease names according to co-occurrence in MEDLINE abstracts, and then ranks the identified terms and selects the genes annotated with the top-ranking terms. In the work of Franke *et al.* [20], a functional human genetic network was developed that integrates information from KEGG, BIND, Reactome, human protein reference database, Gene Ontology, predicted-protein interaction, human yeast two-hybrid interactions, and microrray coexpressions. Gene prioritization is performed by assessing whether genes are close together within the connected gene network. Endeavour [3] takes a machine learning approach by building a model on a training set, then that model is used to rank the test set of candidate genes according to the similarity to the model. The similarity is computed as the correlation for vector space data and BLAST score for sequence data. Endeavour incorporates multiple genomic data sources (microarray, InterPro, BIND, sequence, GO annotation, Motif, Kegg, EST, and text mining) and builds a model on each source of individual prioritization results. Finally, these results are combined through order statistics to a final score that offers an insight on how related a candidate gene to the training genes on the basis of information from multiple knowledge sources. More recently, CAESAR [22] has been developed as a text mining based gene prioritization tool for complex traits. CAESAR ranks genes by comparing the standard correlation of term-frequency vectors (TF profiles) of annotated terms in different ontological descriptions and integrates multiple ranking results by arithmetical (*min*, *max*, and *average*) and parametric integrations.

To evaluate the prioritization model, genes that are known relevant to the same disease are constructed as a disease-specific training set. A prioritization model is first built on this training set, then that model is used to rank a test set of candidate genes according to their similarity to the model. The performance is evaluated by checking the positions of the real relevant genes in the ranking of a test set. A perfect prioritization should rank the gene with the highest relevance to the biomedical concept, represented by the training set, at the highest position (at the top). The interval between the real position of that gene with the top is similar to the error. For a prioritization model, minimizing this error is equal to improving the ranking position of the most relevant gene and in turn it reduces the number of irrelevant genes to be investigated in biological experimental validation. So a model with smaller error is more efficient and accurate to find disease relevant genes and that error is also used as a performance indicator for model comparison [63]. The ranking of candidate genes is usually based on scores. Assuming larger score represents higher similarity towards the prioritization model, in benchmark study, one can label the real relevant genes as class "+1" and other irrelevant genes as class "-1" and plot the Receiver operating characteristic (ROC) curves to compare different models by the values of area under the ROC curve (AUC). The error of prioritization is thus equivalent to 1 minus the AUC value.

## 5.3 Background: Clustering by Heterogeneous Data Sources

Clustering by multiple (heterogeneous) data sources is an ongoing topic with many interests. Recently, Wolf *et al.* [60] have investigated the memory persistence (long term or short term memory) of bacteria by observing a strain of Bacillus subtilis at different experimental conditions and developmental times. These multiple observations are then analyzed by clustering to quantify the mutual information the bacterium "remembers" at different stages. Some other approaches address consensus clustering to combine multiple partitions generated on a single dataset, for instance, the analysis of microarray data by Monti *et al.* [37] and Yu *et al.* [65]. Asur *et al.* [4] adopt consensus clustering methods to combine matrices generated by 3 different types of measurements (topological measure, mutual information measure and GO annotations) to cluster Protein-Protein Interaction networks. Lange and Buhmann [30] merge similarity matrices and phrase multi-source clustering as a non-negative matrix factorization problem. The kernel fusion problem for clustering is connected to many active works in machine learning and optimization, for instance, the framework of linear kernel fusion for binary supervised learning task proposed by Lanckriet *et al.* [29] and Bach *et al.* [7] and its extension to multi-classes problem proposed by Ye *et al.* [62]. Sonnenburg *et al.* simplify the computational burden of kernel fusion by Semi-infinite programming (SIP) [47]. On the basis of kernel fusion, Chen *et al.* [11] propose a clustering algorithm called nonlinear adaptive distance metric learning as an analogue of Lanckriet et al.'s statistical framework for clustering. Yu *et al.* [64] propose a clustering algorithm, OKKC, for heterogeneous data fusion and combine text mining data and bibliometrics data to explore the structure mapping of journal sets [32]. In this Chapter, we systematically evaluate and compare 12 representative algorithms from two main approaches, ensemble clustering and kernel fusion, to combine the multi-view data. Our experimental result shows that Ward linkage, OKKC, and EACAL perform better than other methods. The number of disease genes in our benchmark data is imbalanced, which may partially affect the evaluation of clustering results.

## 5.4 Single View Gene Prioritization: A Fragile Model with Respect to the Uncertainty

In our previous work [63], we found the *configuration* of text mining model has a strong impact on the quality of prioritization model. The term *configuration* denotes the triplet choice of domain vocabulary, representation scheme, and ranking algorithm in text mining based gene prioritization. According to the result of full benchmark experiments shown in Table 5.1, the improperly selected *configuration* could lead to a large error (no-voc, single max, TFIDF, 0.3757) on prioritization, which is more than 7 times larger than the error of a carefully selected *configuration* (eVOC, 1-SVM, IDF, 0.0477). If the prioritization result is used as the reference list in biological validation, the difference of efficiency gained from a good *configuration* and a bad *configuration* will be remarkable. However, how to determine the good *configuration* for prioritization is a non-trivial issue.

## 5.5 Data Fusion for Gene Prioritization: Distribution Free Method

In Endeavour [3, 55], the combination of multi-source prioritization is based on a generic model named order statistics. A Q statistics is calculated from all rank

**Table 5.1** Errors of Leave-one-out validation of prioritization on single text mining profiles. The values in this table are obtained by leave-one-out validation of 618 disease relevant genes in 9,999 random candidate genes. The validations are repeated 10 times for all *configurations* and the deviations are all smaller than 0.0001. For each column, which means comparing the performance of algorithms using the same vocabulary and representation scheme, the smallest error is shown in **Bold**, the largest error is shown in *Italic*. The introduction of experimental setup is available in [63].

| | | GO | | EVOC | | MESH | | LDDB | | OMIM | | ALLVOC | |
|---|---|---|---|---|---|---|---|---|---|---|---|---|---|
| No. | Algorithm | *idf* | *tfidf* | *idf* | *tfidf* | *idf* | *tfidf* | *idf* | *tfidf* | *idf* | *tfidf* | *idf* | *tfidf* |
| 1 | 1-SVM | **0.0758** | **0.0916** | **0.0477** | **0.1316** | **0.0497** | 0.1107 | **0.0877** | 0.1713 | **0.0714** | 0.1288 | 0.0936 | 0.1745 |
| 2 | Correlation | 0.0781 | 0.1168 | 0.0499 | 0.137 | 0.0553 | **0.1103** | 0.089 | 0.1928 | 0.0798 | 0.1364 | **0.0905** | 0.1736 |
| 3 | 1-NN | 0.1203 | 0.1201 | 0.1004 | 0.1676 | 0.1274 | 0.1150 | 0.1332 | *0.2832* | 0.1034 | 0.1296 | 0.1249 | 0.1816 |
| 4 | 2-NN $\gamma$ | 0.1038 | 0.1138 | 0.082 | 0.1513 | 0.1073 | 0.1075 | 0.1173 | 0.2570 | 0.0931 | **0.1243** | 0.1155 | 0.1760 |
| 5 | 2-NN $\delta$ | 0.1297 | 0.1203 | 0.0994 | 0.1596 | 0.1421 | 0.1122 | 0.1359 | 0.2572 | 0.1156 | 0.1278 | 0.1288 | 0.1767 |
| 6 | 2-NN $\kappa$ | 0.0971 | 0.1152 | 0.0733 | 0.1471 | 0.0960 | 0.1088 | 0.1111 | 0.2511 | 0.0933 | 0.1333 | 0.1167 | 0.1792 |
| 7 | 3-NN $\gamma$ | 0.0973 | 0.1113 | 0.0744 | 0.1473 | 0.0974 | 0.1047 | 0.1104 | 0.2498 | 0.1103 | 0.2360 | 0.1110 | 0.1731 |
| 8 | 3-NN $\delta$ | 0.1313 | 0.1162 | 0.0961 | 0.1495 | 0.1394 | 0.1122 | 0.1365 | 0.2458 | 0.1174 | 0.1254 | 0.1146 | 0.1824 |
| 9 | 3-NN $\kappa$ | 0.0970 | 0.1162 | 0.0670 | 0.1460 | 0.0918 | 0.1090 | 0.1044 | 0.2446 | 0.0959 | 0.1341 | 0.1157 | 0.1801 |
| 10 | 2-mean min | 0.0888 | 0.1180 | 0.0641 | 0.1529 | 0.0833 | 0.1094 | 0.0982 | 0.2303 | 0.0899 | 0.1332 | 0.1088 | 0.1776 |
| 11 | 2-mean avg | 0.0822 | 0.1241 | 0.0564 | 0.1611 | 0.0638 | 0.1148 | 0.0885 | 0.1956 | 0.0828 | 0.1483 | 0.0956 | 0.1820 |
| 12 | 2-mean max | 0.1098 | 0.1695 | 0.0812 | 0.1968 | 0.0906 | 0.1619 | 0.1135 | 0.1991 | 0.1134 | 0.1977 | 0.1266 | 0.2387 |
| 13 | 3-mean min | 0.0974 | 0.1202 | 0.0778 | 0.1559 | 0.0987 | 0.1132 | 0.1059 | 0.2572 | 0.0955 | 0.1311 | 0.1184 | 0.1827 |
| 14 | 3-mean avg | 0.0830 | 0.1243 | 0.0558 | 0.1508 | 0.0645 | 0.1156 | 0.0899 | 0.1975 | 0.0850 | 0.1433 | 0.0983 | 0.1833 |
| 15 | 3-mean max | 0.1293 | 0.2019 | 0.0882 | 0.2237 | 0.1065 | 0.2017 | 0.1208 | 0.1849 | 0.1323 | 0.2043 | 0.1556 | 0.2836 |
| 16 | SL min | 0.0858 | 0.1184 | 0.0601 | 0.1560 | 0.0694 | 0.1110 | 0.0922 | 0.2223 | 0.0900 | 0.1381 | 0.0913 | **0.1749** |
| 17 | SL avg | 0.1073 | 0.1659 | 0.0717 | 0.2069 | 0.0866 | 0.1625 | 0.0923 | 0.2031 | 0.1018 | 0.1742 | 0.1079 | 0.2281 |
| 18 | SL max | *0.2888* | *0.3095* | *0.2133* | *0.3212* | *0.2529* | *0.3027* | *0.1823* | **0.1657** | *0.2276* | *0.2756* | 0.2789 | *0.3757* |
| 19 | CL min | 0.0957 | 0.1199 | 0.082 | 0.156 | 0.0888 | 0.1091 | 0.098 | 0.232 | 0.0956 | 0.1337 | 0.1028 | 0.1766 |
| 20 | CL avg | 0.1018 | 0.1443 | 0.0776 | 0.1788 | 0.0871 | 0.1413 | 0.0946 | 0.2000 | 0.0994 | 0.155 | 0.1062 | 0.1995 |
| 21 | CL max | 0.2219 | 0.2262 | 0.1481 | 0.2549 | 0.1766 | 0.2405 | 0.1665 | 0.1906 | 0.1774 | 0.2183 | 0.2239 | 0.2992 |
| 22 | AL min | 0.0884 | 0.119 | 0.0679 | 0.1612 | 0.0749 | 0.1075 | 0.0895 | 0.2189 | 0.0878 | 0.1381 | 0.0961 | 0.1707 |
| 23 | AL avg | 0.1097 | 0.1666 | 0.0783 | 0.2068 | 0.088 | 0.1594 | 0.0886 | 0.195 | 0.1035 | 0.1747 | 0.1107 | 0.2267 |
| 24 | AL max | 0.2827 | 0.3039 | 0.2119 | 0.3173 | 0.2341 | 0.2896 | 0.185 | 0.1557 | 0.2464 | 0.2738 | *0.294* | 0.3736 |
| 25 | WL min | 0.0839 | 0.1176 | 0.0614 | 0.151 | 0.0765 | 0.1129 | 0.0965 | 0.2334 | 0.0889 | 0.1336 | 0.1088 | 0.1774 |
| 26 | WL avg | 0.0795 | 0.1285 | 0.0567 | 0.1571 | 0.0577 | 0.1241 | 0.0896 | 0.2013 | 0.0817 | 0.1483 | 0.0927 | 0.1878 |
| 27 | WL max | 0.1017 | 0.1823 | 0.0934 | 0.2149 | 0.0813 | 0.1754 | 0.1214 | 0.2074 | 0.1164 | 0.2033 | 0.1239 | 0.2564 |

ratios using the joint cumulative distribution of an $N$-dimensional *order statistics* as previously done by Stuart *et al.* [51]:

$$Q(r_1, r_2, ..., r_N) = N! \int_0^{r_1} \int_{s_1}^{r_2} ... \int_{r_{N-1}}^{r_N} ds_N ds_{N-1} ... ds_1 \tag{5.1}$$

To reduce the complexity, Aerts *et al.* implement a much faster alternative formula [3], given by

$$V_k = \sum_{i=1}^{k} (-1)^{i-1} \frac{V_{k-i}}{i!} r_{N-k+1}^{i} \tag{5.2}$$

with $Q(r_1, r_2, ..., r_N) = N!V_N$, $V_0 = 1$, and $r_i$ is the rank ratio for data source $i$. The Q statistics for randomly and uniformly drawn rank ratios is found approximately distributed as a beta distribution when $N \leq 5$, and a gamma distribution for $N > 5$. According to the cumulative distribution, we obtain P-value for every Q statistic from the order statistics [3]. In this way, the original $N$ rankings are combined to a ranking of p-values, as shown in Figure 5.1. The main problem for the generic model based integration is that the performance highly relies on the distribution estimation of the gene prioritization scores. In text mining, the genes are often expressed as sparse vectors spanned in very high dimensional space, therefore, to estimate the distribution of the scores ("distances") among these genes is not always reliable, *e.g.*, "the Pearson variation of the corresponding distance distribution degrades to 0 with increasing dimensionality"[8]. To tackle this problem, one may refine the estimation by designing new distance functions taking the high dimensionality affect into account, such as the work presented by Hsu and Chen [25]. Alternatively, one may adopt the algorithms based on statistical learning theory [58] where the error bound is distribution-free. De Bie *et al.* [15] propose a kernel novelty detection method on the basis of statistical learning theory and prove the error bound of false negative genes in prioritization. Given that bound, the number of false positives is controlled by the total number of positives [15]. To control the number of the false positive, the kernel matrix is centered for the whole genome. In practical, it is centered for the union set of the training genes and the test genes. The algorithm of kernel novelty detection is similar to the $L_\infty$ 1-SVM MKL introduced in (3.15) but involves more statistical learning considerations, given by [15]

$$\begin{aligned} &\min_{\alpha} \; t \\ &\text{s.t.} \;\; t \geq \alpha^T \frac{K_j}{\upsilon_j} \alpha, \; j = 1, ..., p \\ &0 \leq \alpha_i \leq \frac{1}{\nu N}, \; i = 1, ..., N \\ &\sum_{i=1}^{N} \alpha_i = 1, \end{aligned} \tag{5.3}$$

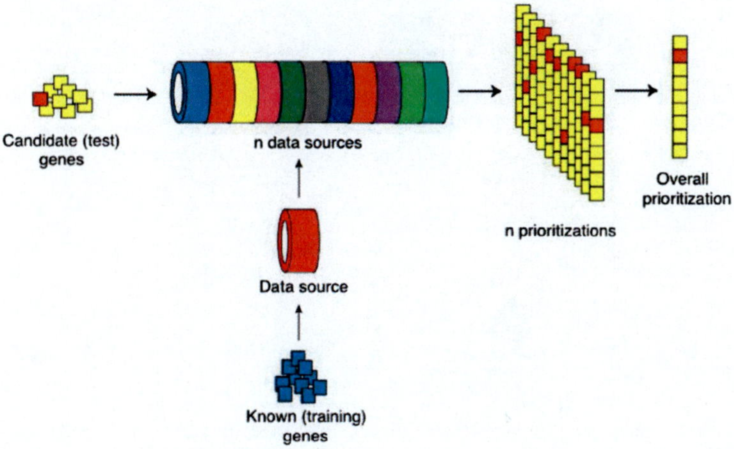


**Fig. 5.1** The conceptual framework of gene prioritization by data fusion [3, 55]. The unified prioritization is obtained by the generic integration of individual prioritization results based on order statistics.

and the prioritization function is given by

$$f(x) = \frac{1}{\sqrt{\alpha^T \Omega \alpha}} \sum_{i=1}^{N} \alpha_i \left[ \sum_{j=1}^{p} \theta_j \frac{K_j(\mathbf{x}, \mathbf{x}_i)}{\upsilon_j} \right], \tag{5.4}$$

where

$$\Omega = \left\{ \sum_{j=1}^{p} \theta_j \frac{K_j}{\upsilon_j} \middle| \sum_{j=1}^{p} \theta_j = 1, \ \forall j, \ \theta_j \geq 0 \right\}. \tag{5.5}$$

In (5.3), (5.4), and (5.5), $N$ is the number of training genes, $p$ is the number of kernels, $\nu$ is the regularization parameter, $\upsilon_j$ is a positive constant controlling the complexity measured from each kernel $K_j$ [29]. Empirically, $\upsilon_j$ is often equated as the trace of the $j$-th centred genome-wide kernel matrix divided by the total number of genes in the genome (practically the sum of training plus test genes), thus, the trace of the combined kernel matrix $\Omega$ of the whole genome is equivalent to the number of genes in the whole genome and the norm $K(\mathbf{x}, \mathbf{x})$ of a gene is equal to 1 on average [15]. The conceptual framework of the 1-SVM MKL approach for gene prioritization is illustrated in Figure 5.2. This approach is empirically compared with the generic model adopted in Endeavour. When using the same genomic data sources, the distribution-free kernel method outperforms the generic model.

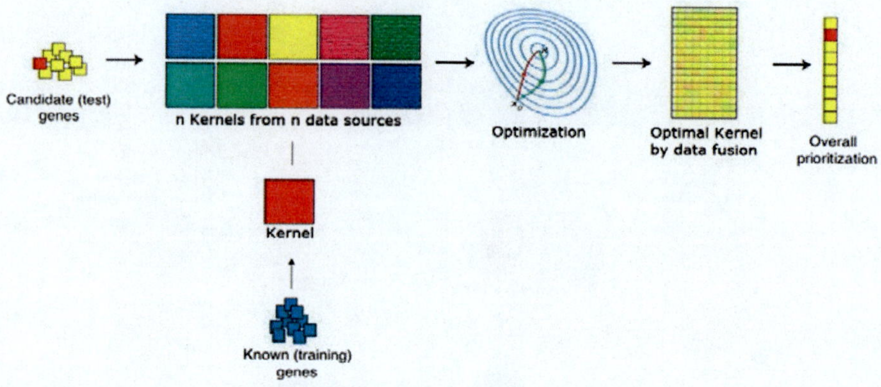


**Fig. 5.2** The conceptual framework of kernel gene prioritization by data fusion proposed in [15]. The unified prioritization is obtained by the scoring function based on the optimal combination of kernel matrices in equation (5.4).

## 5.6 Multi-view Text Mining for Gene Prioritization

### *5.6.1 Construction of Controlled Vocabularies from Multiple Bio-ontologies*

We select vocabularies from nine bio-ontologies for text mining, among which five of them (GO, MeSH, eVOC, OMIM and LDDB) have proven their merit in our earlier work of text based gene prioritization [63] and text based cytogenetic bands mapping [57]. Besides these five, we select four additional ontologies (KO, MPO, SNOMED CT, and UniprotKB) because they are also frequently adopted in the identification of genetic diseases and signaling pathways, for instance, in the works of Gaulton *et al.* [22], Bodenreider [10], Mao *et al.* [34], Smith *et al.* [49], and Melton *et al.* [36]. The nine bio-ontlogies are briefly introduced as follows.

#### The Gene Ontology

GO [14] provides consistent descriptions of gene and gene-product attributes in the form of three structured controlled vocabularies that each provide a specific angle of view (biological processes, cellular components and molecular functions). GO is built and maintained with the explicit goal of applications in text mining and semantic matching in mind [57]. Hence, it is an ideal source as domain-specific views in our approach. We extract all the terms in GO (due to the version released in December, 2008) as the CV of GO.

### Medical Subject Headings

MeSH is a controlled vocabulary produced by NLM for indexing, cataloging, and searching biomedical and health-related information and documents. The descriptors or subject headings of MeSH are arranged in a hierarchy. MeSH covers a broad ranges of topics and its current version consists of 16 top level categories. Though most of the articles in MEDLINE are already annotated with MeSH terms, our text mining process does not rely on these annotations but indexes the MEDLINE repository automatically with the MeSH descriptors (version 2008).

### Online Mendelian Inheritance in Man's Morbid Map

OMIM [35] is a database that catalogues all the known diseases with genetic components. It contains available links between diseases and relevant genes in the human genome and provides references for further research and tools for genomic analysis of a catalogued gene. OMIM is composed of two mappings: the OMIM Gene Map, which presents the cytogenetic locations of genes that are described in OMIM; the OMIM Morbid Map, which is an alphabetical list of diseases described in OMIM and their corresponding cytogenetic locations. Our approach retrieves the disease descriptions from the OMIM Morbid Map (version due to December, 2008) as the CV.

### London Dysmorphology Database

LDDB is a database containing information over 3000 dysmorphic and neurogenetic syndromes, which is initially developed to help experienced dysmorphologists to arrive at the correct diagnosis in difficult cases with multiple congenital anomalies [59]. Information in the database is constantly updated and over 1000 journals are regularly reviewed to ascertain appropriate reports. The London Neurology Database (LNDB) is a database of genetic neurological disorders based on the same data structure and software as the LDDB [6]. We extract the dysmorphology taxonomies from LNDB (version 1.0.11) and select the vocabulary terms.

### eVOC

eVOC [28] is a set of vocabularies that unifies gene expression data by facilitating a link between the genome sequence and expression phenotype information. It was originally categorized as four orthogonal controlled vocabularies (anatomical system, cell type, pathology, and developmental stage) and now extended into 14 orthogonal subsets subsuming the domain of human gene expression data. Our approach selects the vocabulary from the eVOC version 2.9.

### KEGG Orthology

KO is a part of the KEGG suite [27] of resources. KEGG is known as a large pathway database and KO is developed to integrate pathway and genomic information in KEGG. KO is structured as a directed acyclic graph (DAG) hierarchy of four flat levels [34]. The top level consists of the following five categories: metabolism, genetic information processing, environmental information processing, cellular processes and human diseases. The second level divides the five functional categories into finer sub-categories. The third level corresponds directly to the KEGG pathways, and the fourth level consists of the leaf nodes, which are the functional terms. In literature, KO has been used as an alternative controlled vocabulary of GO for automated annotation and pathway identification [34]. The KO based controlled vocabulary in our approach is selected on the version due to December 2008.

### Mammalian Phenotype Ontology

MPO [49] contains annotations of mammalian phenotypes in the context of mutations, quantitative trait loci and strains which was initially used in Mouse Genome Database and Rat Genome Database to represent phenotypic data. Because mouse is the premier model organism for the study of human biology and disease, in the CAESAR [22] system, MPO has also been used as a controlled vocabulary for text mining based gene prioritization of human diseases. The MPO based controlled vocabulary in our approach is selected on the version due to December 2008.

### Systematized Nomenclature of Medicine–Clinical Terms

SNOMED is a huge and comprehensive clinical terminology, originally created by the College of American Pathologists and, now owned, maintained, and distributed by the International Health Terminology Standards Development Organization (IHTSDO). SNOMED is a very "fine-grained" collection of descriptions about care and treatment of patients, covering areas like diseases, operations, treatments, drugs, and healthcare administration. SNOMED has been investigated as an ontological resource for biomedical text mining [10] and also has been used in patient-based similarity metric construction [36]. We select the CV on the SNOMED (version due to December, 2008) obtained from the Unified Medical Language System (UMLS) of NLM.

### Universal Protein Knowledgebase

UniProtKB [18] is a repository for the collection of functional information on proteins with annotations developed by European Bioinformatics Institute (EBI). Annotations in UniProtKB are manually created and combined with non-redundant protein sequence database, which brings together experimental results, computed features and scientific conclusions. Mottaz *et al.* [38] design a mapping procedure to link the UniProt human protein entries and corresponding OMIM entries to the MeSH disease terminology. The vocabulary applied in our approach is selected on UniProt release 14.5 (due to December, 2008).

The terms extracted from these bio-ontologies are stored as bag-of-words and preprocessed for text mining. The preprocessing includes transformation to lower case, segmentation of long phrases, and stemming. After preprocessing, these vocabularies are fed into a Java program based on Apache Java Lucene API to index the titles and abstracts of MEDLINE publications relevant to human genes.

### 5.6.2 Vocabularies Selected from Subsets of Ontologies

As mentioned, in some "fine-grained" bio-ontologies the concepts and terminologies are labeled in multiple hierarchies, denoted as sub-ontologies, to represent domain concepts at various levels of specificity. For instance, GO consists of three sub-ontologies: biological process, cellular component and molecular function. MeSH descriptors are arranged in 16 hierarchical trees at the top level. In SNOMED, the medical terminologies are composed of 19 higher level hierarchies. eVOC ontology contains 14 orthogonal vocabulary subsets, in which the terms contained are strictly non-overlapping. To investigate whether more specific vocabularies can improve the effectiveness of the text mining model, we select terms from the sub-ontologies of GO, MeSH, SNOMED, and eVOC and compose the corresponding subset CVs. Considering the main objective as disease-associated gene identification, only the most relevant sub-ontologies (6 from eVOC, 7 from MeSH and 14 from SNOMED) are selected. To distinguish the gene-by-term profiles obtained from subset CVs with those obtained from complete CVs, we denote the former one as subset CV profile and latter one as complete CV profile.

### 5.6.3 Merging and Mapping of Controlled Vocabularies

The strategy of incorporating multiple CVs may be alternatively achieved by merging terms of several vocabularies together. To investigate this, we merge the terms of all 9 complete CVs as a union of vocabulary and denote the corresponding gene-by-term text mining result as "merge-9 profile". Furthermore, we notice the lexical variants across multiple ontologies: a concept may be represented as different terms due to the diversities of professional expressions. For example, the MeSH term "denticles" is expressed as "dental pulp stones" in OMIM and as "pulp stone" in SNOMED. To resolve these variants, we refer the UMLS Metathesaurus to map terminological variants as unified concepts. We download the concept names and sources file (MRCONSO.RRF) from UMLS Metathesaurus, which provides the mapping of atoms (each occurrence of unique string or concept name within each source vocabulary) to unified concepts. In text mining, we build a synonym vocabulary to map and aggregate the occurrences of various synonym terms are mapped and aggregated as the indices of the unified concept. The obtained results are gene-by-concept profile which the features are the unique and permanent concept identifiers defined by UMLS Metathesaurus. Among the nine vocabularies adopted in our approach, only four of them (GO, MeSH, OMIM and SNOMED) are included in UMLS and resolved in the MRCONSO.RRF file. Therefore, to fairly compare the effect of concept mapping, we also create "merge-4" gene-by-term profile using the union of the four vocabularies in indexing. Then, we map the lexical variants as concepts and the result is denoted as "concept-4" profile. Moreover, to create a naive baseline, we also index the MEDLINE corpus without using any controlled vocabulary. All the terms appeared in the corpus are segmented as single words[1] and the results are expressed by these vast amount of words, denoted as "no-voc profile".

[1] We didn't consider the phrases of multiple words because there would be immense amount of combinations.

**Table 5.2** Overview of the controlled vocabularies applied in the multi-view approach. The *Number of indexed terms* of controlled vocabularies reported in this table are counted on indexing results of human related publications only so their numbers are smaller than those in our earlier work [63], which were counted on all species appeared in GeneRIF. The *Number of terms in CV* are counted on the vocabularies independent to the indexing process.

| No. | CV | Nr. terms CV | Nr. indexed terms |
|---|---|---|---|
| 1 | **eVOC** | 1659 | 1286 |
| 2 | eVOC anatomical system | 518 | 401 |
| 3 | eVOC cell type | 191 | 82 |
| 4 | eVOC human development | 658 | 469 |
| 5 | eVOC mouse development | 369 | 298 |
| 6 | eVOC pathology | 199 | 166 |
| 7 | eVOC treatment | 62 | 46 |
| 8 | **GO** | 37069 | 7403 |
| 9 | GO biological process | 20470 | 4400 |
| 10 | GO cellular component | 3724 | 1571 |
| 11 | GO molecular function | 15282 | 3323 |
| 12 | **KO** | 1514 | 554 |
| 13 | **LDDB** | 935 | 890 |
| 14 | **MeSH** | 29709 | 15569 |
| 15 | MeSH analytical | 3967 | 2404 |
| 16 | MeSH anatomy | 2467 | 1884 |
| 17 | MeSH biological | 2781 | 2079 |
| 18 | MeSH chemical | 11824 | 6401 |
| 19 | MeSH disease | 6717 | 4001 |
| 20 | MeSH organisms | 4586 | 1575 |
| 21 | MeSH psychiatry | 1463 | 907 |
| 22 | **MPO** | 9232 | 3446 |
| 23 | **OMIM** | 5021 | 3402 |
| 24 | **SNOMED** | 311839 | 27381 |
| 25 | SNOMED assessment scale | 1881 | 810 |
| 26 | SNOMED body structure | 30156 | 2865 |
| 27 | SNOMED cell | 1224 | 346 |
| 28 | SNOMED cell structure | 890 | 498 |
| 29 | SNOMED disorder | 97956 | 13059 |
| 30 | SNOMED finding | 51159 | 3967 |
| 31 | SNOMED morphologic abnormality | 6903 | 2806 |
| 32 | SNOMED observable entity | 11927 | 3119 |
| 33 | SNOMED procedure | 69976 | 9575 |
| 34 | SNOMED product | 23054 | 1542 |
| 35 | SNOMED regime therapy | 5362 | 1814 |
| 36 | SNOMED situation | 9303 | 2833 |
| 37 | SNOMED specimen | 1948 | 742 |
| 38 | SNOMED substance | 33065 | 8948 |
| 39 | **Uniprot** | 1618 | 520 |
| 40 | Merge-9 | 372527[2] | 50687 |
| 41 | Merge-4 | 363321 | 48326 |
| 42 | Concept-4 | 1420118 | 44714 |
| 43 | No-voc | - | 259815 |

**Table 5.3** The number of overlapping terms in different vocabularies and indexed terms. The upper triangle matrix shows the numbers of overlapping terms among vocabularies independent to indexing. The lower triangle matrix shows the numbers of overlapping indexed terms. The second horizontal row (from the top) are the numbers of the terms in vocabularies independent to indexing. The second vertical column (from the left) are the numbers of the indexed terms.

| | | eVOC | GO | KO | LDDB | MeSH | MPO | OMIM | SNOMED | Uniprot |
|---|---|---|---|---|---|---|---|---|---|---|
| | | 1659 | 37069 | 1514 | 935 | 29709 | 9232 | 5021 | 311839 | 1618 |
| eVOC | 1286 | - | 370 | 16 | 118 | 827 | 566 | 325 | 876 | 46 |
| GO | 7403 | 358 | - | 404 | 74 | 3380 | 1234 | 659 | 4772 | 325 |
| KO | 554 | 16 | 344 | - | 1 | 383 | 72 | 120 | 489 | 44 |
| LDDB | 890 | 118 | 74 | 1 | - | 346 | 275 | 205 | 498 | 16 |
| MeSH | 15569 | 784 | 2875 | 344 | 343 | - | 2118 | 1683 | 12483 | 373 |
| MPO | 3446 | 554 | 1177 | 72 | 271 | 2007 | - | 823 | 2729 | 146 |
| OMIM | 3402 | 322 | 655 | 119 | 205 | 1644 | 816 | - | 2275 | 161 |
| SNOMED | 27381 | 814 | 3144 | 380 | 492 | 8900 | 2508 | 2170 | - | 593 |
| Uniprot | 520 | 46 | 301 | 42 | 16 | 361 | 146 | 157 | 371 | - |

### 5.6.4 *Text Mining*

We refer the mapping of genes and publications in Entrez GeneRIF and index a subset of MEDLINE literature repository (as of 10 December, 2008) that consists of 290,000 human gene-related publications. In the first step the MEDLINE documents are indexed and the doc-by-term (or doc-by-concept) vectors are constructed. In the second step, we averagely combine the document-by-term (document-by-concept) vectors as gene-by-term (gene-by-concept) vectors according to the GeneRIF mapping. The detail of the text mining process is presented in our earlier work [23, 63]. Table 5.2 lists all the CVs applied in our approach. Table 5.3 illustrates the overlapping terms among the nine complete CVs.

Preliminary result shows that the weighting scheme of terms also influences the performance of gene-by-term data in biological validation [63]. When the same vocabulary and the ranking algorithm are applied in prioritization, the IDF representation generally outperforms the TF-IDF and the binary representations. Therefore, in this article all the term profiles are represented in the IDF weighting scheme.

### 5.6.5 *Dimensionality Reduction of Gene-By-Term Data by Latent Semantic Indexing*

We have introduced the subset CVs method to reduce the number of terms expressing the genes. Alternatively, we also apply *Latent semantic indexing* (LSI) to reduce the number of term features. On the one hand, the information expressed on vast numbers of terms is mapped to a smaller number of latent factors so the irrelevant information is reduced. On the other hand, we expect that LSI does not compromise the information required for prioritization and clustering. In implementation, we use the Matlab function *svds* to solve the eigenvalue problem of the sparse gene-by-term matrix of the whole human genome (22,743 human genes). To calculate the total variance on this huge matrix is very computational expensive, so we sort the eigenvalues obtained by the sparse eigenvalue decomposition. To determine the number of latent factors, we select the dimension where the corresponding smallest eigenvalue is less than 0.05% of the sum of all eigenvalues.

## 5.6.6 Algorithms and Evaluation of Gene Prioritization Task

The methods to combine models for prioritization are roughly classified as two approaches: ensemble of rankings and fusion of sources (kernels).

### 5.6.6.1 Ensemble Ranking

In ensemble ranking, the prioritization is first carried on each individual model and then multiple ranking results are combined. Since our main objective is to integrate the models, we use the same algorithm, standard correlation, as the base method to obtain ranking results on individual models. Using other algorithms, the results after model integration may not be significantly different and the computational complexity is more likely to be higher than the standard correlation algorithm.

The ranking results is integrated either as ranking orders (ratios) or as ranking scores. To compare them, we implement three integration algorithms. Two of them are basic operators to calculate the average or maximal value of multiple ranking scores. The third one is based on order statistics as defined in equation (5.1).

### 5.6.6.2 Kernel Fusion for Prioritization

The kernel fusion method for gene prioritization is based on the 1-SVM MKL problem defined in (5.3). We apply this 1-SVM method to combine kernels derived from the multi-view data for gene prioritization. Because the dimensionality of gene-by-term profile is high, we only use linear function to construct the kernel matrices. One of the main features of the 1-SVM is the sparsity of its solutions, which the dominant coefficient may be assigned on one or two kernel matrices. This property is useful to distinguish a small amount of relevant sources from large amount of irrelevant sources in data fusion. However, in biomedical application, the data sources are usually preprocessed and have high relevances w.r.t. the problem. Sparse solution may be too selective, in this case, to thoroughly combine the redundant and complementary information in these data sources. To balance the effect of sparse coefficients (most of $\theta_j$ are equal to 0) and non-sparse coefficients in model generalization, we try 3 different values of the regularization parameter $\theta_{min}$ to restrict the optimization process as a lowerbound of coefficient assigned on each kernel. When $\theta_{min} = 0$, there is no lowerbound and the optimization procedure will probably result in the sparse coefficients. When $\theta_{min} = 0.5/N$, each kernel is insured to have a minimum contribution in data fusion. When $\theta_{min} = 1/N$, the kernel matrices are averagely combined. We also try the $L_2$ 1-SVM MKL introduced in chapter 3. In our implementation, the QCLP problem in (5.3) is solved by SeDuMi 1.2 [48].

#### 5.6.6.3 Evaluation of Prioritization

The prioritization result is evaluated by leave-one-out (LOO) method [63]. In each experiment, given a disease gene set which contains $K$ genes, one gene, termed the "defector" gene, is deleted from the set of training genes and added to 99 randomly selected test genes (test set). We use the remained $K-1$ genes (training set) to build our prioritization model. Then, we prioritize the test set which contains 100 genes by the trained model and determine the ranking of that defector gene in test data. The prioritization performance is thus equivalent to the error (1 minus the AUC value).

### *5.6.7 Benchmark Data Set of Disease Genes*

We validate the clustering results with the human disease benchmark data set of Endeavour [3], which consists of 620 relevant genes from 29 diseases. Genes from the same disease are constructed as a disease-specific training set used to evaluate the prioritization and clustering performance. For prioritization, we perform 620 rankings (with each gene left out once) on 99 candidate human genes randomly selected from the human genomic. The prioritization is repeated 20 times (with randomly permuted 99 random genes each time for each left out gene) and the average Error value is reported as the final result.

## 5.7 Results of Multi-view Prioritization

### *5.7.1 Multi-view Performs Better than Single View*

According to the experimental result, integration of the *multi-view* data obtains significantly better performance than the best individual data. Among the different approaches we tried, the best performance is obtained by combining 9 CV profiles as kernels in the LSI reduced dimensionality (1-SVM+LSI, Error of AUC = 0.0335). This error is only a half of the best single CV profile (LDDB, 0.0792). The ROC curves of leave-one-out performance of different approaches are presented in Figure 5.3. Without LSI, 1-SVM data fusion reduces the error from 0.0792 to 0.0453. By coupling LSI with 1-SVM, the error is further reduced from 0.0453 to 0.0335. Considering the cost and effort of validating the false positive genes in lab experiments, the improvement from 0.0792 to 0.0335 is quite meaningful because it means that when prioritizing 100 candidate genes, our proposed method can save the effort of validating 4 false positive genes. The obtained result is also comparable to the performance of the existing systems. In the Endeavour system [3], the same disease gene benchmark dataset and the same evaluation method is implemented. Endeavour is different from our approach mainly in two aspects. Firstly, Endeavour combines one textual data (GO-IDF profile obtained from free literature text mining) with nine other biological data sources. Also, there is no dimensionality reduction used in it. Secondly, Endeavour applies order statics to integrate the models. The performance obtained in our paper is much better than Endeavour (Error=0.0833). Moreover, the performance is also better than the result obtained by De Bie *et al.* [15]. In their approach, they use the same data sources as Endeavour and apply the 1-SVM MKL algorithm for model integration (best performance Error=0.0477, $\theta_{min} = 0.5/k$). Since the methods and the disease gene benchmark data are exactly the same, the improvement can only be attributed to the multi-view text mining and the LSI dimensionality reduction. It is also notice that the $L_2$-norm 1-SVM MKL performs better than the $L_\infty$ ($\theta_{min} = 0$) approach. When optimizing the $L_2$-norm, the integration of 9 CVs has the error of 0.0587, and the integration of 9 LSI profiles has the error of 0.0392. This result is consistent with our hypothesis that non-sparse kernel fusion may perform better than sparse kernel fusion. However, the best result in the multi-view prioritization is obtained by $L_1$ ($\theta_{min} = 1$) approach.

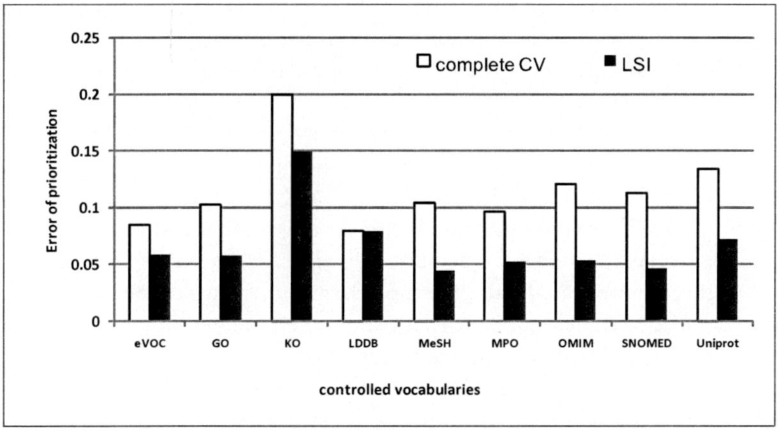


**Fig. 5.3** Prioritization results obtained by complete CV profiles and LSI profiles

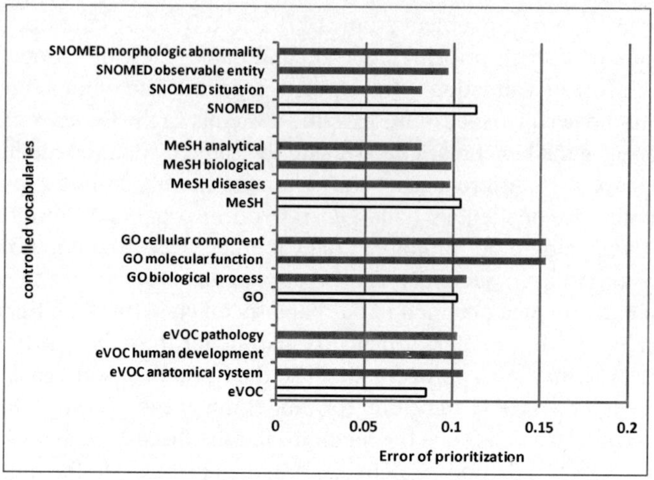


**Fig. 5.4** Prioritization results obtained by complete CV profiles and subset CV profiles

## 5.7.2 *Effectiveness of Multi-view Demonstrated on Various Number of Views*

To further demonstrate the effectiveness of multi-view text mining, we evaluated the performance on various number of views. The number was increased from 2 to 9 and three different strategies were adopted to add the views. Firstly, we simulated a random strategy by enumerating all the combinations of views from the number of 2 to 9. The combinations of 2 out of 9 views is $C_9^2$, 3 out of 9 is $C_9^3$, and so on. We calculated the average performance of all combinations for each number of views. In the second and the third experiment, the views were added by two different heuristic rules. We ranked the performance of the nine views from high to low was LDDB, eVOC, MPO, GO, MeSH, SNOMED, OMIM, Uniprot, and KO. The second strategy combined best views first and increases the number from 2 to 9. In the third strategy, the irrelevant views were integrated first. The results obtained by these three strategies are presented in Figure 5.6. The performance of the random strategy increases steadily with the number of views involved in integration. In the best view first strategy, the performance increased and reached the ideal performance, then started to decrease when more irrelevant views are involved. The ideal performance of prioritization was obtained by combining the five best views (Error of AUC = 0.0431) by the 1-SVM method applied on averagely combined kernel. The generic integration method (order statistic) did not perform well on high dimensional gene-by-term data. The performance of integrating all CVs was comparable to the ideal performance, which shows that the proposed multi-view approach is quite robust to the irrelevant views. Furthermore, the merit in practical explorative analysis is that the near-optimal result can be obtained without evaluating each individual model. In the third strategy, because the combination starts from the irrelevant views first, the performance was not comparable to the random or the ideal case. Nevertheless, as

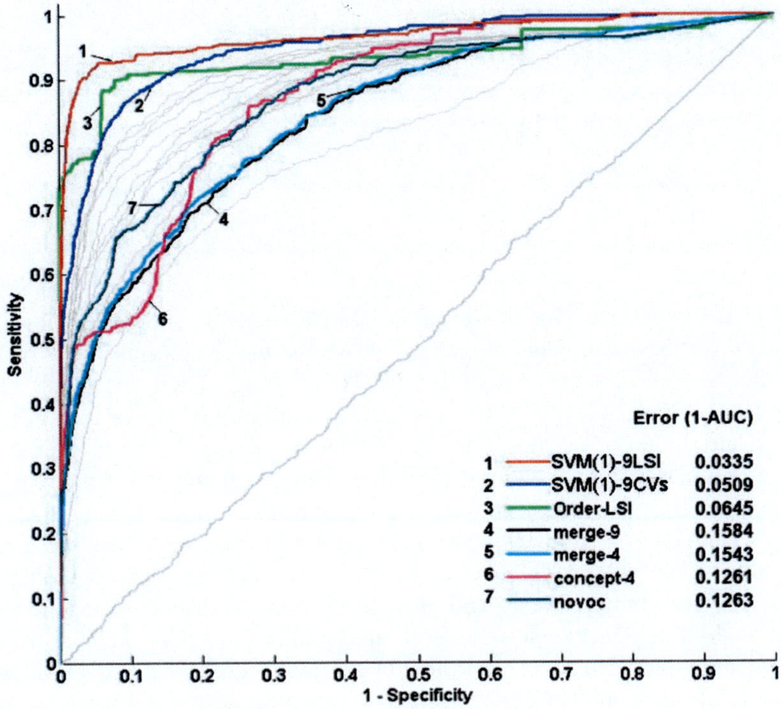


**Fig. 5.5** ROC curves of prioritization obtained by various integration methods. The light grey curves represent individual textual data. The near-diagonal curve is obtained by prioritization of random genes.

shown, the performance of the multi-view approach was always better than the best single view involved in integration. Collectively, this experiment clearly illustrated that the *multi-view* approach is a promising and reliable strategy for disease gene prioritization.

### 5.7.3 Effectiveness of Multi-view Demonstrated on Disease Examples

To explain why the improvements take place when combining multiple views, we show an example taken from prioritization of MTM1, a gene relevant to the disease Myopathy. In the disease benchmark data set, Myopathy contains 41 relevant genes so we build the disease model by using the other 40 genes and leave MTM1 out with 99 random selected genes for validation. In order to compare the rankings,

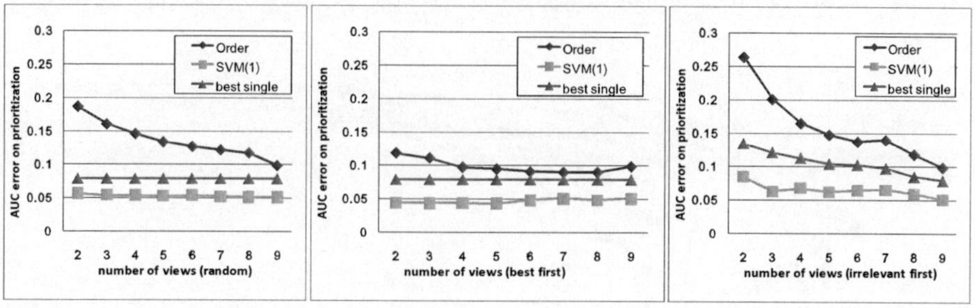


**Fig. 5.6** Multi-view prioritization with various number of views

only in the experiment for this example, the 99 random candidate genes are kept identical for different views. In Table 5.4, we list the ranking positions of MTM1 and the false positive genes in all 9 CVs. When using LDDB vocabulary, 3 "false positive genes" (C3orf1, HDAC4, and CNTFR) are ranked higher than MTM1. To investigate the terms causing this, we sort the correlation score of each term between the disease model and the candidate genes. It seems that C3orf1 is ranked at the top mainly because of the high correlation with terms "skeletal, muscle, and heart". HDAC4 is ranked at the second position because of terms "muscle, heart, calcium, and growth". CTNFR is ranked at the third position due to the terms like "muscle, heart, muscle weak, and growth". As for the real disease relevant gene MTM1, the high correlated terms are "muscle, muscle weak, skeletal, hypotonia, growth, and lipid". However, according to our knowledge, none of these three genes (C3orf1, HDAC4, and CNTFR) is actually known to cause any disease. Escarceller *et al.* [19] show that C3orf1 seems to be enhanced in heart and skeletal muscle, but there is no clue that it has directly relation with the disease. HDAC4 is found in the work of Little *et al.* "as a specific downstream sbstrate of CaMKIIdeltaB in cardiac cells and have broad applications for the signaling pathways leading to cardiac hypertrophy and heart failure" [31]. In the papers of Glenisson *et al.* [24] and Cohen *et al.* [13], HDAC4 is found to have a role in muscle, which means it might be a good candidate but has not yet been proved directly related to the disease. For CNTFR, it has been found that in heterozygotic mice inactivation of the CNTFR leads to a slight muscle weakness [41]. In the papers of Roth *et al.* [44] and De Mars *et al.* [16], CNTFR is shown related to muscular strength in human. Collectively, although these 3 genes found by LDDB vocabulary have not yet been reported as direct disease causing factors, the prioritization result is still meaningful because in literature they have many similar correlated terms with the disease model as the real disease causing gene does. Especially, according to literature, HDAC4 and CNFR seem to be nice candidates to muscular disorder. Though LDDB ranks 3 "false positive" genes higher than the real disease relevant gene, eVOC and GO rank the left out gene MTM1 as the top candidate gene. In eVOC, the most important correlated terms are "muscle, sever, disorder, ...". In GO, the terms are "muscle, mutation, family, sever, ...". When combining multi-view data for prioritization, the ranking of LDDB is complemented by eVOC and GO thus MTM1 is ranked as the top gene.

**Table 5.4** Prioritization results of MTM1 by different CV profiles

| CV | Rank | Gene | correlated terms |
|---|---|---|---|
| LDDB | 1 | C3orf1 | muscle, heart, skeletal |
| | 2 | HDAC4 | muscle, heart, calcium, growth |
| | 3 | CNTFR | muscle, heart, muscle weak, growth |
| | **4** | **MTM1** | muscle, muscle weak, skeletal, hypotonia, growth, lipid |
| eVOC | **1** | **MTM1** | muscle, sever, disorder, affect, human, recess |
| MPO | 1 | HDAC4 | muscle, interact, protein, domain, complex |
| | 2 | HYAL1 | sequence, human, protein, gener |
| | 3 | WTAP | protein, human, sequence, specif |
| | 4 | FUT3 | sequence, alpha, human |
| | ... | | |
| | **15** | **MTM1** | myopathy, muscle, link, sequence, disease, sever |
| GO | 1 | **MTM1** | muscle, mutate, family, gene, link, seqeuence, sever |
| MeSH | 1 | HYAL1 | human, protein, clone, sequence |
| | 2 | LUC7L2 | protein, large, human, function |
| | **3** | **MTM1** | myopathy, muscle, mutate, family, gene, missens |
| SNOMED | 1 | S100A8 | protein, large, human, function |
| | 2 | LUC7L2 | protein, large, human, function |
| | 3 | LGALS3 | human, protein, express, bind |
| | ... | | |
| | **23** | **MTM1** | muscle, mutate, family, gene, link |
| OMIM | 1 | HDAC4 | muscle, interact, protein, bind |
| | 2 | MAFK | sequence, protein, gene, asthma relat trait |
| | 3 | LUC7L2 | protein, large, function, sequence |
| | 4 | SRP9L1 | sequence, protein, length, function |
| | ... | | |
| | **50** | **MTM1** | muscle, family, gene, link, sequence, disease, sever, weak |
| Uniprot | 1 | **MTM1** | gene, protein, function |
| KO | 1 | S100A8 | protein, bind, complex, specif, associ, relat |
| | 2 | PRF1 | specif, protein, contain, activ |
| | ... | | |
| | **56** | **MTM1** | protein, large, specif, contain |
| Multi-view | **1** | **MTM1** | |
| | 2 | HDAC4 | |
| | 3 | CNTFR | |

## 5.8 Multi-view Text Mining for Gene Clustering

We investigate two fundamental tasks in disease associated gene research: prioritization and clustering. Both tasks have attracted lots of efforts in the literature, whereas their definitions and interpretations may vary by approach. On one hand, computational gene prioritization methods *rank* the large amount of candidate disease genes according to their likeliness of being involved in a certain disease. On the other hand, clustering analysis explores the disease-gene associations by partitioning the genes based on the experimental findings described in the scientific literature. These two tasks basically share a similar assumption: In prioritization, the similarity among genes associated to the same disease is assumed to be higher than the similarity with random genes. In the case of multiple diseases, the problem can also be formulated as a clustering problem. The assumption is that the similarity of genes relevant to the same disease (within-disease-cluster similarity) is higher than the similarity of genes relevant to different diseases (between-disease-cluster similarity). Thus, we expect these genes to demonstrate some "natural partitions" according to the type of diseases. Therefore, we are able to evaluate the performance of prioritization task and the clustering task using the same disease benchmark data.

### *5.8.1 Algorithms and Evaluation of Gene Clustering Task*

**Ensemble clustering**

In ensemble clustering, we apply $k$-means clustering using Euclidean distance as the "base clustering algorithm" on a single data source to obtain the partition; then we combine the multiple partitions as a consolidate partition via consensus functions. We also tried other candidate algorithms (*e.g.*, hierarchical clustering, self-organizing maps) and other distance measures (*e.g.*, Mahalanobis distance, Minkowski distance), although we observe some discrepancies of performance on individual gene-by-term data, the difference after multi-view integration is not significant. In literature, various consensus functions have been proposed for ensemble clustering. We select 6 popular ones and compare them in our approach.

**CSPA, HGPA, and MCLA:** Strehl and Ghosh [50] formulate the optimal consensus as the partition that shares the most information with the partitions to combine, as measured by the Average Normalized Mutual Information. They use three heuristic consensus algorithms based on graph partitioning, called Cluster based Similarity Partition Algorithm (CSPA), Hyper Graph Partitioning Algorithm (HGPA) and Meta Clustering Algorithm (MCLA) to obtain the combined partition.

**QMI:** Topchy *et al.* [54] formulate the combination of partitions as a categorical clustering problem. In their approach, a category utility function is adopted to evaluate the quality of a "median partition" as a summary of the ensemble. They prove that maximizing this category utility function implies the same clustering ensemble criterion as maximizing the generalized mutual information based on quadratic entropy (QMI). Furthermore, the maximization of the category utility function is equivalent to the square error based clustering criterion when the number of clusters is fixed. The final consensus partition is obtained by applying the $k$-means algorithm on the feature space transformed by the category utility function.

**EACAL:** Fred and Jain [21] introduce the concept of Evidence Accumulation Clustering (EAC) that maps the individual data partitions as a clustering ensemble by constructing a co-association matrix. The entries of the co-association matrix are interpreted as votes on the pairwise co-occurrences of objects, which is computed as the number of occurrences each pair of objects appears in the same cluster of an individual partition. Then the final consensus partition is obtained by applying single linkage (SL) and average linkage (AL) algorithms on the co-association matrix. According to their experiments, average linkage performs better than single linkage so in this paper we apply Evidence Accumulation Clustering with average linkage (EACAL) as the representative algorithm for comparison.

**AdacVote:** Ayad and Kamel [5] propose an Adaptive cumulative Voting (AdacVote) method to compute an empirical probability distribution summarizing the ensemble. The goal of this ensemble is to minimize the average squared distance between the mapped partitions and the combined partition. The cumulative voting method seeks an adaptive reference partition and incrementally updates it by averaging other partitions to relax the dependence of the combined partition on the selected reference. In the AdacVote they proposed, the partitions are combined in the decreasing order of their entropies.

### Kernel fusion for clustering

An alternative approach to combine multi-view data for clustering is achieved by fusing the similarity matrices [30], as known as the *kernel fusion* approach. Kernel fusion integrates data before clustering (early integration), whereas ensemble clustering aggregates partitions after clustering (late integration). We implement 5 kernel fusion algorithms and cross-compare their performance. In the present paper, the kernel matrices are all constructed by linear functions because the text mining data is in very high dimension.

**Hierarchical clustering:** We average the kernels of multi-view data and transform it into a distance matrix in Hilbert space, given by [46]:

$$d_{\phi}(x,z) = \langle \phi(x), \phi(x) \rangle - 2\langle \phi(x), \phi(z) \rangle + \langle \phi(z), \phi(z) \rangle. \tag{5.6}$$

When the kernel mapping $\phi(\cdot)$ is based on a linear function and $x$ and $z$ are data vectors normalized by the norm, $d_{\phi}(x,z)$ boils down to the Euclidean distance between $x$ and $z$. When $\phi(\cdot)$ is based on nonlinear mapping (*e.g.*, RBF functions,

Polynomial functions), the distance $d_\phi(x,z)$ does not have direct interpretation in the original space of $x$ and $z$. Given the distance matrix, we can apply linkage methods (i.e., single linkage, complete linkage, average linkage, and ward linkage) and obtain the hierarchical clustering result in Hilbert space.

**OKKC:** The optimized data fusion for kernel $k$-means clustering (OKKC) is proposed in the previous chapter.

#### Evaluation of clustering

As explained before, we assess the clustering performance biologically by labeled disease benchmark data. Two external validations, Rand Index (RI) [42] and Normalized Mutual Information (NMI) [50], are applied and their definitions are given as follows. Given a set of $N$ genes $X = \{x_1,...,x_N\}$ and two partitions to compare, $\mathscr{C} = \{c_1,...,c_N\}$ and $\mathscr{L} = \{l_1,...,l_N\}$. In our problem, $\mathscr{C}$ and $\mathscr{L}$ are respectively the cluster indicators and the disease labels of the set of genes $X$. We refer that (1) $a$, the number of pairs of genes in $X$ that are in the same set in $\mathscr{C}$ and in the same set in $\mathscr{P}$; (2) $b$, the number of pairs of genes in $X$ that are in different sets in $\mathscr{C}$ and in different sets in $\mathscr{P}$; (3) $c$, the number of pairs of genes in $X$ that are in the same set in $\mathscr{C}$ and in different sets in $\mathscr{P}$; (4) $d$, the number of pairs of genes in $X$ that are in different sets in $\mathscr{C}$ and in the same set in $\mathscr{P}$.

RI is defined as

$$RI = \frac{a+b}{a+b+c+d} \ . \tag{5.7}$$

For binary class problem, the RI value ranges from 0 to 1 and the value of random partitions is 0.5.

NMI is defined as

$$NMI = \frac{2 \times M(\mathscr{C},\mathscr{P})}{E(\mathscr{C})E(\mathscr{P})} \ , \tag{5.8}$$

where $M(\mathscr{P},\mathscr{C})$ is the mutual information between the indicators, $E(\mathscr{C})$ and $E(\mathscr{P})$ are the entropies of the indicators and the labels. For a balanced clustering problem, if the indicators and the labels are independent, the NMI value approaches 0.

### 5.8.2 *Benchmark Data Set of Disease Genes*

We validate the clustering results using the same benchmark data set we use in prioritization. The schema of the clustering evaluation is depicted in Figure 5.7. For each time, the relevant genes of each paired diseases combination are selected and clustered into two groups, then the performance is evaluated using the disease labels. The genes which are relevant to both diseases in the paired combination are removed before clustering (totally less then 5% genes have been removed). Finally, the average performance of all the 406 paired combinations is used as the overall clustering performance. The tasks on all 406 paired combinations are repeated 20 times and the mean value of RI and NMI of all tasks in all repetitions is reported as the final result.

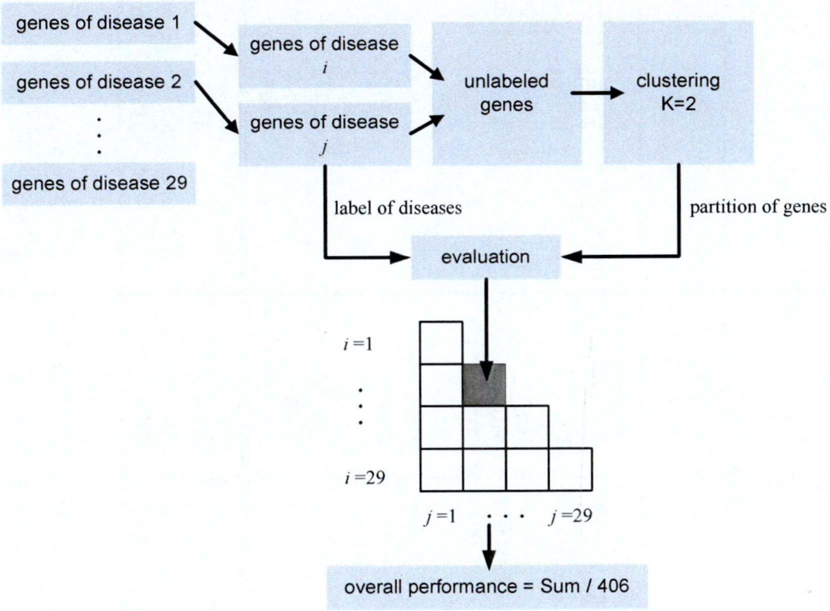


**Fig. 5.7** Conceptual scheme of clustering disease relevant genes

## 5.9 Results of Multi-view Clustering

### *5.9.1 Multi-view Performs Better than Single View*

In clustering, as shown in Table 5.5, three algorithms obtained significant performance. Ward linkage applied on average combination of kernels showed the best performance (RI=0.8236, NMI=0.6015). EACAL (RI=0.7741, NMI=0.5542) and OKKC without regularization ($\mu_{min}$=0, RI=0.7641, NMI=0.5395) also performed better than the best single view data (LDDB, RI=0.7586, NMI=0.5290).

**Table 5.5** Clustering performance obtained by the single controlled vocabulary and the multi-view approach. For integration of 9 LSI and 35 subset CVs, only the best three results are shown.

| Single CV | RI | NMI | Integration (9 CVs) | RI | NMI |
|---|---|---|---|---|---|
| LDDB | **0.7586** ± 0.0032 | **0.5290**± 0.0032 | Ward linkage | **0.8236** ± 0 | **0.6015** ± 0 |
| OMIM | 0.7216 ± 0.0009 | 0.4606 ± 0.0028 | EACAL | 0.7741 ± 0.0041 | 0.5542 ± 0.0068 |
| Uniprot | 0.7130 ± 0.0013 | 0.4333 ± 0.0091 | OKKC($\mu_{min}$=0) | 0.7641 ± 0.0078 | 0.5395 ± 0.0147 |
| eVOC | 0.7015 ± 0.0043 | 0.4280 ± 0.0079 | MCLA | 0.7596 ± 0.0021 | 0.5268 ± 0.0087 |
| MPO | 0.7064 ± 0.0016 | 0.4301 ± 0.0049 | QMI | 0.7458 ± 0.0039 | 0.5084 ± 0.0063 |
| MeSH | 0.6673 ± 0.0055 | 0.3547 ± 0.0097 | OKKC($\mu_{min}$=1/N) | 0.7314 ± 0.0054 | 0.4723 ± 0.0097 |
| SNOMED | 0.6539 ± 0.0063 | 0.3259 ± 0.0096 | AdacVote | 0.7300 ± 0.0045 | 0.4093 ± 0.0100 |
| GO | 0.6525 ± 0.0063 | 0.3254 ± 0.0092 | CSPA | 0.7011 ± 0.0065 | 0.4479 ± 0.0097 |
| KO | 0.5900 ± 0.0014 | 0.1928 ± 0.0042 | Complete linkage | 0.6874 ± 0 | 0.3028 ± 0 |
| | | | Average linkage | 0.6722 ± 0 | 0.2590 ± 0 |
| | | | HGPA | 0.6245 ± 0.0035 | 0.3015 ± 0.0071 |
| | | | Single linkage | 0.5960 ± 0 | 0.1078 ± 0 |
| Integration (9 LSI) | RI | NMI | Integration (35 subset CVs) | RI | NMI |
| Ward linkage | 0.7991 ± 0 | 0.5997 ± 0 | Ward linkage | 0.8172 ± 0 | 0.5890 ± 0 |
| OKKC($\mu_{min}$=0) | 0.7501 ± 0.0071 | 0.5220 ± 0.0104 | OKKC($\mu_{min}$=0) | 0.7947 ± 0.0052 | 0.5732 ± 0.0096 |
| EACAL | 0.7511 ± 0.0037 | 0.5232 ± 0.0075 | EACAL | 0.7815 ± 0.0064 | 0.5701 ± 0.0082 |

We apply paired t-test to evaluate the statistical significance of the data fusion performance. As shown in Table 5.6, the three presented data fusion approaches significantly improve the performance over the best individual data.

**Table 5.6** Statistical analysis of the performance obtained on disease relevant genes clustering. Performance of WL, EACAL, and OKKC is compared with the best single data source. The $p$ values are obtained by paired t-test on 20 random repetitions.

| | RI | | | NMI | | |
|---|---|---|---|---|---|---|
| | mean | std. | $p$ value | mean | std. | $p$ value |
| Best single data | 0.7586 | 0.0032 | - | 0.5290 | 0.0032 | - |
| WL | 0.8236 | 0 | $7.97 \times 10^{-47}$ | 0.6015 | 0 | $6.45 \times 10^{-46}$ |
| EACAL | 0.7741 | 0.0041 | $3.39 \times 10^{-31}$ | 0.5542 | 0.0068 | $1.69 \times 10^{-33}$ |
| OKKC | 0.7641 | 0.0078 | $1.32 \times 10^{-5}$ | 0.5395 | 0.0147 | $1.89 \times 10^{-5}$ |

## 5.9.2 *Dimensionality Reduction of Gene-By-Term Profiles for Clustering*

The same LSI profiles and subset CV profiles for prioritization are also used in clustering task. As shown in Figure 5.8 and Figure 5.9, some LSI profiles were slightly better, others were slightly worse than the the complete profiles. Some subset CVs performed better than the complete CV. In particular, SNOMED situation and MeSH diseases outperformed significantly the complete CVs.

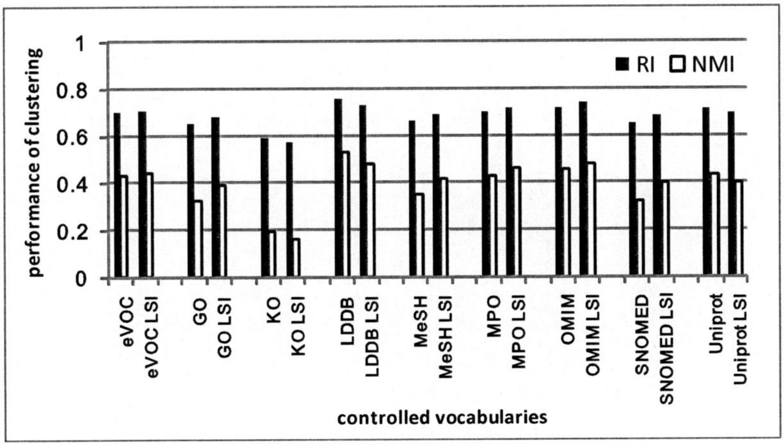


**Fig. 5.8** Clustering results obtained by complete CV and LSI profiles

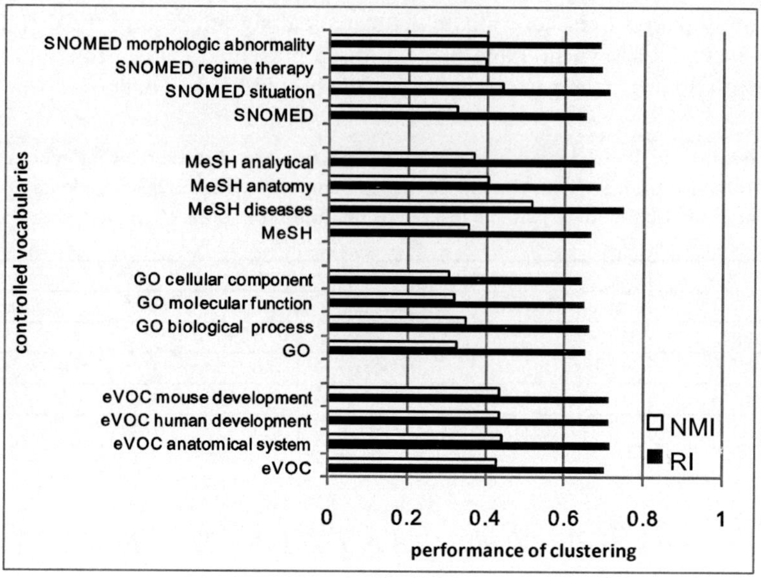


**Fig. 5.9** Clustering results obtained by complete CV and subset CV profiles

Analogue to the prioritization task, we integrated 9 complete CVs, 9 LSI profiles, and 35 subset CVs for clustering and evaluated the performance. As shown in Table 5.6 and Figure 5.10, the best result was obtained by combing 9 complete CVs with Ward linkage, OKKC ($\mu_{min}$=0), and EACAL. Other comparing methods did not obtain better results than the best single CV.

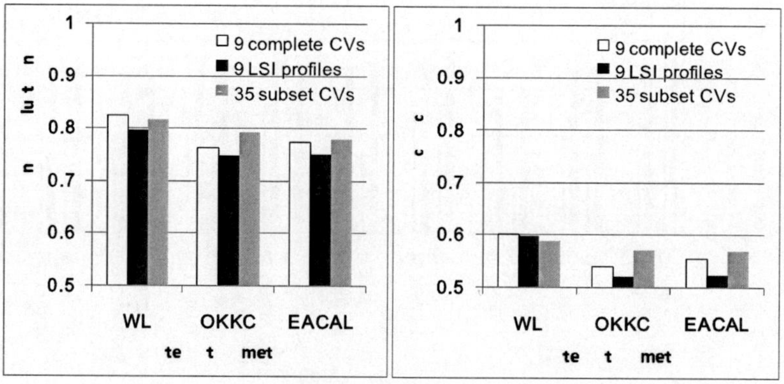


**Fig. 5.10** Clustering results obtained by multi-view data integration

### 5.9.3 Multi-view Approach Is Better than Merging Vocabularies

The *multi-view* approach also performed than merging vocabularies. As shown in Table 5.7, the best performance of merging vocabularies (RI=0.6333, NMI=0.2867) was significantly lower than the multi-view result obtained by Ward linkage (RI=0.8236, 0.6015).

**Table 5.7** Clustering performance obtained by merging controlled vocabularies, concept mapping and no vocabulary indexing. The merge-9, merge-4 and concept-4 profiles were clustered by *k*-means in 20 random repetitions and the mean values and deviations of evaluations are shown in the table. The novoc profile was only evaluated once by *k*-means because of the extremely high dimension and the computational burden.

| Merging vocabulary | RI | NMI |
|---|---|---|
| merge-9 | $0.6321 \pm 0.0038$ | $0.2830 \pm 0.0079$ |
| merge-4 | $0.6333 \pm 0.0053$ | $0.2867 \pm 0.0085$ |
| concept-4 | $0.6241 \pm 0.0056$ | $0.2644 \pm 0.0111$ |
| novoc | 0.5630 | 0.0892 |

### 5.9.4 Effectiveness of Multi-view Demonstrated on Various Numbers of Views

We also evaluated the performance of clustering on various numbers of views. The ideal performance of clustering was obtained by combining the four best views (RI=0.8540, NMI=0.6644) using the ward linkage method. As shown in Figure 5.11, the general trends were similar to our discussion in the prioritization task.

### 5.9.5 Effectiveness of Multi-view Demonstrated on Disease Examples

We present an example of clustering genes relevant to two diseases: breast cancer and muscular dystrophy. Each disease contains 24 relevant genes and there is no overlapping gene among them. We list the confusion tables and the mis-partitioned genes of each individual view in Table 5.8. As it is shown, all individual views produce several mis-partitioned genes. When these views are combined and clustered by the ward linkage method, all the genes are correctly partitioned by their disease labels.

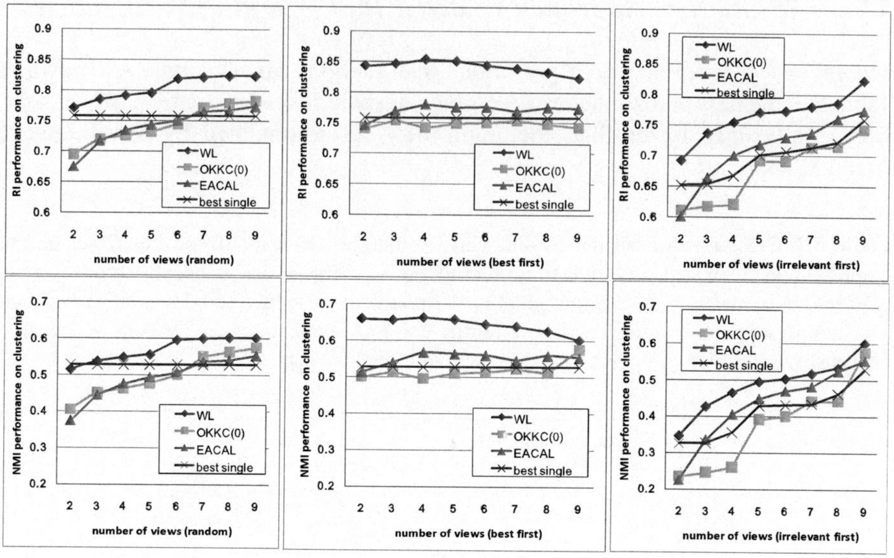


**Fig. 5.11** Multi-view clustering with various number of views

**Table 5.8** Clustering breast cancer and muscular dystrophy relevant genes by different CVs and the multi-view approach

| CV | B. Cancer | M. Dystrophy | mis-partitioned genes |
|---|---|---|---|
| LDDB | 22 | 2 | RP11-49G10.8, FKTN |
| | 0 | 24 | |
| eVOC | 23 | 1 | RP11-49G10.8, FKTN |
| | 7 | 17 | LMNA, COL6A1, MYF6, CHEK2, SGCD, FKRP, DMD |
| MPO | 23 | 1 | RP11-49G10.8 |
| | 1 | 23 | SGCD |
| GO | 23 | 1 | RP11-49G10.8 |
| | 7 | 17 | LMNA, COL6A1, MYF6, CHEK2, SGCD, FKRP, DMD |
| MeSH | 23 | 1 | RP11-49G10.8 |
| | 2 | 22 | SGCD,COL6A3 |
| SNOMED | 24 | 0 | |
| | 6 | 18 | LMNA,COL6A1,MYF6,TRIM32,SGCD,DMD |
| OMIM | 24 | 0 | |
| | 1 | 23 | SGCD |
| Uniprot | 24 | 0 | |
| | 4 | 20 | MYF6,CHEK2,SGCD,FKRP |
| KO | 19 | 5 | SLC22A18, RP11-49G10.8, FKTN, PABPN1, CAPN3 |
| | 6 | 18 | PPM1D, MYF6, SGCD, FKRP, COL6A3, DYSF |
| Multi-view (WL) | 24 | 0 | |
| | 0 | 24 | |

## 5.10 Discussions

The strategy of model integration has already been used in several text mining applications. The concept of *multi-view* is proposed by Bickel and Scheffer [9] in web document clustering analysis to combine intrinsic view of web pages (text based similarity) and extrinsic view (citation link based similarity). In our paper, we borrow the term *multi-view* in the text mining context as gene profiles represented by different CVs. Neveol *et al.* [39] combine three different methods (dictionary lookup, post-processing rules and NLP rules) to identify MeSH main head-ing/subheading pairs from medical text. Chun *et al.* [12] develop an integrative system to extract disease-gene relations from MEDLINE. Jimeno *et al.* [26] combine three methods (dictionary look-up, statistical scoring, and MetaMap) to recognize disease names on a corpus of annotated sentences. Gaulton *et al.* [22] adopt 3 different ontologies and 8 data sources in the CAESAR system to annotate human genes as disease associated candidates. When annotating multiple data sources with different relevant terms from ontologies, each gene may get multiple scores of relevance with the input text query. CAESAR combines the scores using 4 basic methods: maximum, sum, normalized average, and a transformed score penalized by the number of genes annotated in a given data source. Our approach is different from CAESAR by exploiting all the relevant MEDLINE abstracts for indexing so the gene textual profiles are retrieved from vast amounts of gene-based free-text information available in the literature. Yamakawa *et al.* [61] combine 3 different sources (GO, Locuslink and HomoloGene) to create gene list annotated by GO terms. Then, a decomposition method (ETMIC situation decomposition) is applied to extract the multi-aspect information from the target gene list, resulting in several bipartite graphs describing the relationships between small subset of genes and GO terms. Their approach shares the same motivation as ours in obtaining more refined characteristics of genes separated in different aspects (views). However, their approach has not shown how multi-aspect gene annotations can improve the process of biomedical knowledge discovery. The main limitation of combining the LSI with data fusion is that the latent factors cannot be easily interpreted so it is hard to investigate the important terms in prioritization. The idea of *multi-view* text mining is not restricted to the selection of CVs. For example, instead of using the curated GeneRIF as the mapping of genes to publications, one can detect gene names expressed in the text automatically by natural language processing (NLP) and create new gene profiles according to this new mapping. One can also retrieve the literature co-occurences of genes and produce new *view* about the gene-to-gene interactions. Combining these views will probably lead to new insight about the relationship between diseases and genes.

The interpretation of text based prioritization is limited by LSI, whose latent factors cannot be easily attributed to the terms affecting the prioritization. When combining multi-view data by *k*-means and ensemble algorithms (individual partition created by *k*-means), to estimate the optimal cluster numbers is also difficult because the number of clusters is predefined. The statistical evaluations of clustering quality which is used to indicate the optimal cluster number on single data set are not always reliable for data fusion because they may differ in heterogeneous data sources. To circumvent this problem, one may relax the $k$-means clustering as a spectral clustering [66] thus the optimal cluster number can be investigated from the eigenspectrum. To estimate the optimal cluster number in hierarchical clustering is easier, because it can be estimated by checking the dendrogram. Another limitation in our clustering approach is the ignorance of overlapping genes despite of the fact that a gene may be biologically relevant to several topics (*e.g.*, diseases, functions, processes). Therefore, how to apply "soft clustering" techniques to obtain partitions containing overlapping genes will be the main topic of our future work. The notion of *multi-view* text mining has the potential of incorporating models varied by other parameter. For example, instead of using curated GeneRIF as the mapping of genes to publications, one can detect gene names expressed in the text automatically by natural language processing (NLP) and create new gene profiles according to this mapping. One can also retrieve the relationships of genes from literature, or refer to interaction networks and produce new view specified about relationships of genes. Combining these views will undoubtedly lead to significant and thorough insight about the associations between diseases and genes.

## 5.11 Summary

We have presented the approach of combining *multi-view* text mining models to obtain precise identification of disease relevant genes. These views were specified by multiple controlled vocabularies derived from different bio-ontologies. Using these vocabularies, we have indexed the MEDLINE titles and abstracts relevant to GeneRIF and have obtained a series of gene-by-term profiles. To demonstrate the effectiveness of our approach, we have combined these profiles and evaluated them on two fundamental problems: gene prioritization and clustering. Experimental results have shown that the performance obtained on the *multi-view* approach is significantly better than the single-view data. Nonetheless, the selection of the appropriate integration algorithm was nontrivial. We have cross-compared 4 algorithms in prioritization and 12 algorithms in clustering on a disease benchmark data set containing 29 diseases. In prioritization, the combination of the 1-SVM with LSI performed the best; in clustering, the ward linkage applied on the uniform combination of kernels performed better than other methods.

Second, we have integrated dimensionality reduction of individual data source in the data fusion framework. To tackle the very high dimensionality of text mining data, we have applied LSI, a popular reduction technique in information retrieval, on gene-by-term profiles. Alternatively, we have also pruned the vocabularies according to the hierarchical structures of the bio-ontologies where they were derived. In this way, the gene-by-term profiles specified by a complete CV have been further separated as several subset CV profiles. In some experiments, the LSI and the subset CV profiles have obtained better performance than the complete CV.

Third, we have substantiated the rationale of the proposed "integration after splitting" by comparing three other methods such as vocabulary integration, concept mapping, and no vocabulary indexing. Experiments and validation results have clearly indicated that the proposed *multi-view* approach is a promising strategy.

## References


1. Adie, E.A., Adams, R.R., Evans, K.L., Porteous, D.J., Pickard, B.S.: SUSPECTS: enabling fast and effective prioritization of positional candidates. Bioinformatics 22, 773–774 (2006)
2. Adie, E.A., Adams, R.R., Evans, K.L., Porteous, D.J., Pickard, B.S.: Speeding disease gene discovery by sequence based candidate prioritization. BMC Bioinformatics 6, 55 (2005)
3. Aerts, S., Lambrechts, D., Maity, S., Van Loo, P., Coessens, B., De Smet, F., Tranchevent, L.-C., De Moor, B., Marynen, P., Hassan, B., Carmeliet, P., Moreau, Y.: Gene prioritization through genomic data fusion. Nature Biotechnology 24, 537–544 (2006)
4. Asur, S., Parthasarathy, S., Ucar, D.: An ensemble framework for clustering protein-protein interaction network. Bioinformatics 23, i29–i40 (2007)
5. Ayad, H.G., Kamel, M.S.: Cumulative voting consensus method for partitions with a variable number of clusters. IEEE Trans. PAMI 30, 160–173 (2008)
6. Aymè, S.: Bridging the gap between molecular genetics and metabolic medicine: access to genetic information. European Journal of Pediatrics 159, S183–S185 (2000)
7. Bach, F.R., Lanckriet, G.R.G., Jordan, M.I.: Multiple kernel learning, conic duality, and the SMO algorithm. In: Proceedings of 21st International Conference of Machine Learning. ACM Press, New York (2004)
8. Beyer, K., Goldstein, J., Ramakrishnan, R., Shaft, U.: What is the Nearest Neighbor in High Dimensional Spaces? In: Beeri, C., Bruneman, P. (eds.) ICDT 1999. LNCS, vol. 1540, pp. 217–235. Springer, Heidelberg (1999)
9. Bickel, S., Scheffer, T.: Multi-View Clustering. In: Proc. of IEEE data mining Conference, pp. 19–26. IEEE, Los Alamitos (2004)
10. Bodenreider, O.: Lexical, Terminological, and Ontological Resources for Biological Text Mining. In: Ananiadou, S., McNaught, J.N. (eds.) Text mining for biology and biomedicine, pp. 43–66. Artech House, Boston (2006)
11. Chen, J.H., Zhao, Z., Ye, J.P., Liu, H.: Nonlinear adaptive distance metric learning for clustering. In: Proceeding of ACM KDD, pp. 123–132. ACM Press, New York (2007)
12. Chun, H.W., Yoshimasa, T., Kim, J.D., Rie, S., Naoki, N., Teruyoshi, H.: Extraction of gene-disease relations from MEDLINE using domain dictionaries and machine learning. In: Proceeding of PSB 2006, pp. 4–15 (2007)
13. Cohen, T.J., Barrientos, T., Hartman, Z.C., Garvey, S.M., Cox, G.A., Yao, T.P.: The deacetylase HDAC4 controls myocyte enchancing factor-2-dependent structural gene expression in response to neural activity. The FASEB Journal 23, 99–106 (2009)
14. Consortium: Gene Ontology: Gene ontology: tool for the unification of biology. Nature Genetics 25, 25–29 (2000)
15. De Bie, T., Tranchevent, L.C., Van Oeffelen, L., Moreau, Y.: Kernel-based data fusion for gene prioritization. Bioinformatics 23, i125–i123 (2007)
16. De Mars, G., Windelinckx, A., Beunen, G., Delecluse, G., Lefevre, J., Thomis, M.A.: Polymorphisms in the CNTF and CNTF receptor genes are associated with muscle strength in men and women. Journal of Applied Physiology 102, 1824–1831 (2007)

17. van Driel, M.A., Cuelenaere, K., Kemmeren, P.P.C.W., Leunissen, J.A.M., Brunner, H.G., Vriend, G.: GeneSeeker: extraction and integration of human disease-related information from web-based genetic databases. Nucleic Acids Research 33, 758–761 (2005)
18. Emmert, D.B., Stoehr, P.J., Stoesser, G., Cameron, G.N.: The European Bioinformatics Institute (EBI) databases. Nucleic Acids Research 26, 3445–3449 (1994)
19. Escarceller, M., Pluvinet, R., Sumoy, L., Estivill, X.: Identification and expression analysis of C3orf1, a novel human gene homologous to the Drosophila RP140-upstream gene. DNA Seq. 11, 335–338 (2000)
20. Franke, L., van Bakel, H., Fokkens, L., de Jong, E.D., Egmont-Petersen, M., Wijmenga, C.: Reconstruction of a functional human gene network, with an application for prioritizing positional candidate genes. Am. J. Hum. Genet. 78, 1011–1025 (2006)
21. Fred, A.L.N., Jain, A.K.: Combining multiple clusterings using evidence accumulation. IEEE Trans. PAMI 27, 835–850 (2005)
22. Gaulton, K.J., Mohlke, K.L., Vision, T.J.: A computational system to select candidate genes for complex human traits. Bioinformatics 23, 1132–1140 (2007)
23. Glenisson, P., Coessens, B., Van Vooren, S., Mathys, J., Moreau, Y., De Moor, B.: TXTGate: profiling gene groups with text-based information. Genome Biology 5, R43 (2004)
24. Glenisson, W., Castronovo, V., Waltregny, D.: Histone deacetylase 4 is required for TGFbeta1-induced myofibroblastic differentiation. Biochim. Biophys. Acta 1773, 1572–1582 (2007)
25. Hsu, C.M., Chen, M.S.: On the Design and Applicability of Distance Functions in High-Dimensional Data Space. IEEE Trans. on Knolwedge and Data Engineering 21, 523–536 (2009)
26. Jimeno, A., Jimenez-Ruiz, E., Lee, V., Gaudan, S., Berlanga, R., Rebholz-Schuhmann, D.: Assessment of disease named entity recognition on a corpus of annotated sentences. BMC Bioinformatics 9, S3 (2008)
27. Kanehisa, M., Goto, S., Kawashima, S., Nakaya, A.: The KEGG databases at GenomeNet. Nucleic Acids Research 30, 42–46 (2002)
28. Kelso, J., Visagie, J., Theiler, G., Christoffels, A., Bardien-Kruger, S., Smedley, D., Otgaar, D., Greyling, G., Jongeneel, V., McCarthy, M., Hide, T., Hide, W.: eVOC: A Controlled Vocabulary for Gene Expression Data. Genome Research 13, 1222–1230 (2003)
29. Lanckriet, G.R.G., Cristianini, N., Bartlett, P., Ghaoui, L.E., Jordan, M.I.: Learning the Kernel Matrix with Semidefinite Programming. Journal of Machine Learning Research 5, 27–72 (2004)
30. Lange, T., Buhmann, J.M.: Fusion of Similarity Data in Clustering. Advances in Neural Information Processing Systems 18, 723–730 (2006)
31. Little, G.H., Bai, Y., Williams, T., Poizat, C.: Nuclear calcium/calmodulin-dependent protein kinase IIdelta preferentially transmits signals to histone deacetylase 4 in cardiac cells. The Journal of Biological Chemistry 282, 7219–7231 (2007)
32. Liu, X., Yu, S., Moreau, Y., De Moor, B., Glänzel, W., Janssens, F.: Hybrid Clustering of Text Mining and Bibliometrics Applied to Journal Sets. In: Proc. of the SIAM Data Mining Conference 2009. SIAM Press, Philadelphia (2009)
33. Lopez-Bigas, N., Ouzounis, C.A.: Genome-wide indentification of genes likely to be involved in human genetic disease. Nucleic Acids Research 32, 3108–3114 (2004)
34. Mao, X.: Automated genome annotation and pathway identification using the KEGG Orthology (KO) as a controlled vocabulary. Bioinformatics 21, 3787–3793 (2005)
35. McKusick, V.A.: Mendelian Inheritance in Man. A Catalog of Human Genes and Genetic Disorders, 12th edn. Johns Hopkins University Press, Baltimore (1998)

36. Melton, G.B.: Inter-patient distance metrics using SNOMED CT defining relationships. Journal of Biomedical Informatics 39, 697–705 (2006)
37. Monti, S.: Consensus Clustering: A Resampling-Based Method for Class Discovery and Visualization of Gene Expression Microarray Data. Machine Learning 52, 91–118 (2003)
38. Mottaz, A.: Mapping proteins to disease terminologies: from UniProt to MeSH. BMC Bioinformatics 9, S5 (2008)
39. Neveol, A.: Multiple approaches to fine-grained indexing of the biomedical literature. In: Proceeding of PSB 2007, pp. 292–303 (2007)
40. Perez-Iratxeta, C., Wjst, M., Bork, P., Andrade, M.A.: G2D: a tool for mining genes associated with disease. BMC Genetics 6, 45 (2005)
41. Plun-Favreau, H., Elson, G., Chabbert, M., Froger, J., de Lapeyriére, O., Leliévre, E., Guillet, C., Hermann, J., Gauchat, J.F., Gascan, H., Chevalier, S.: The ciliary neurotrophic factor receptor $\alpha$ component induces the secretion of and is required for functional responses to cardiotrophin-like cytokine. EMBO Journal 20, 1692–1703 (2001)
42. Rand, W.M.: Objective criteria for the evaluation of clustering methods. Journal of the American Statistical Association (American Statistical Association) 66, 846–850 (1971)
43. Risch, N.J.: Searching for genetic determinants in the new millennium. Nature 405, 847–856 (2000)
44. Roth, S.M., Metter, E.J., Lee, M.R., Hurley, B.F., Ferrell, R.E.: C174T polymorphism in the CNTF receptor gene is associated with fat-free mass in men and women. Journal of Applied Physiology 95, 1425–1430 (2003)
45. Shatkay, H., Feldman, R.: Mining the biomedical literature in the genomic era: An overview. Journal of Computational Biology 10, 821–855 (2003)
46. Shawe-Taylor, J., Cristianin, N.: Kernel Methods for Pattern Analysis. Cambridge University Press, Cambridge (2004)
47. Sonnenburg, S., Rätsch, G., Schäfer, C., Schölkopf, B.: Large scale multiple kernel learning. Journal of Machine Learning Research 7, 1531–1565 (2006)
48. Sturm, J.F.: Using SeDuMi 1.02, a MATLAB toolbox for optimization over symmetric cones. Optimization Methods and Software 11/12, 625–653 (1999)
49. Smith, C.L.: The Mammalian Phenotype Ontology as a tool for annotating, analyzing and comparing phenotypic information. Genome Biology 6, R7 (2004)
50. Strehl, A., Ghosh, J.: Clustering Ensembles: a knowledge reuse framework for combining multiple partitions. Journal of Machine Learning Research 3, 583–617 (2002)
51. Stuart, J.M.: A gene-coexpression network for global discovery of con-served genetic modules. Science 302, 249–255 (2003)
52. Tiffin, N.: Prioritization of candidate disease genes for metabolic syndrome by computational analysis of its defining phenotypes. Physiol. Genomics 35, 55–64 (2008)
53. Tiffin, N., Kelso, J.F., Powell, A.R., Pan, H., Bajic, V.B., Hide, W.: Integration of text- and data-mining using ontologies successfully selects disease gene candidates. Nucleic Acids Res. 33, 1544–1552 (2005)
54. Topchy, A.: Clustering Ensembles: models of consensus and weak partitions. IEEE Trans. PAMI 27, 1866–1881 (2005)
55. Tranchevent, L., Barriot, R., Yu, S., Van Vooren, S., Van Loo, P., Coessens, B., De Moor, B., Aerts, S., Moreau, Y.: ENDEAVOUR update: a web resource for gene prioritization in multiple species. Nucleic Acids Research 36, W377–W384 (2008)
56. Turner, F.S., Clutterbuck, D.R., Semple, C.A.M.: POCUS: mining genomic sequence annotation to predict disease genes. Genome Biology 4, R75 (2003)
57. Van Vooren, S., Thienpont, B., Menten, B., Speleman, F., De Moor, B., Vermeesch, J.R., Moreau, Y.: Mapping Biomedical Concepts onto the Human Genome by Mining Literature on Chromosomal Aberrations. Nucleic Acids Research 35, 2533–2543 (2007)

58. Vapnik, V.: Statistical Learning Theory. Wiley Interscience, New York (1998)
59. Winter, R.M., Baraitser, M., Douglas, J.M.: A computerised data base for the diagnosis of rare dysmorphic syndromes. Journal of Medical Genetics 21, 121–123 (1984)
60. Wolf, D.M., Bodin, L.F., Bischofs, I., Price, G., Keasling, J., Arkin, A.P.: Memory in Microbes: Quantifying History-Dependent Behavior in a Bacterium. PLOSone 3, e1700 (2008)
61. Yamakawa, H.: Multi-aspect gene relation analysis. In: Proceeding of PSB 2005, pp. 233–244 (2005)
62. Ye, J.P., Ji, S.W., Chen, J.H.: Multi-class Discriminant Kernel Learning via Convex Programming. Jounral of Machine Learning Research 9, 719–758 (2008)
63. Yu, S., Tranchevent, L.-C., Van Vooren, S., De Moor, B., Moreau, Y.: Comparison of vocabularies, representations and ranking algorithms for gene prioritization by text mining. Bioinformatics 24, i119–i125 (2008)
64. Yu, S., Tranchevent, L.-C., Liu, X., Glänzel, W., Suykens, J.A.K., De Moor, B., Moreau, Y.: Optimized data fusion for kernel K-means clustering. Internal Report 08-200, ESAT-SISTA, K.U.Leuven, Lirias number: 242275 (2008) (submitted for publication)
65. Yu, Z.W., Wong, H.-S., Wang, H.Q.: Graph-based consensus clustering for class discovery from gene expression data. Bioinformatics 23, 2888–2896 (2007)
66. Zha, H., Ding, C., Gu, M., He, X., Simon, H.: Spectral relaxation for K-means clustering. Advances in Neural Information Processing Systems 13, 1057–1064 (2001)

# Chapter 6
# Optimized Data Fusion for $k$-means Laplacian Clustering

## 6.1 Introduction

Clustering is a fundamental problem in unsupervised learning and a number of different algorithms and methods have emerged over the years. $k$-means and spectral clustering are two popular methods for clustering analysis. $k$-means (KM) is proposed to cluster attribute-based data into $k$ numbers of clusters with the minimal distortion [4, 8]. Another well known method, spectral clustering (SC) [18, 20], is also widely adopted in many applications. Unlike KM, SC is specifically developed for graphs, where the data samples are represented as vertices connected by non-negatively weighted undirected edges. The problem of clustering on graphs belongs to another paradigm than the algorithms based on the distortion measure. The goal of graph clustering is to find partitions on the graph such that the edges between different groups have a very low weight [31]. To model this, different objective functions are adopted and the typical criteria include the RatioCut [11], the normalized cut [20], and many others. To solve these objectives, the discrete constraint of the clustering indicators is usually relaxed to real values; thus, the approximated solution of spectral clustering can be obtained from the eigenspectrum of the graph Laplacian matrix. Many investigations (*e.g.*, [6]) have shown the connection between KM and SC. Moreover, in practical applications, the weighted similarity matrix is often used interchangeably as the kernel matrix in KM or the adjacency matrix in SC.

Recently, a new algorithm, kernel Laplacian clustering (KL), is proposed to combine a kernel and a Laplacian simultaneously in clustering analysis [25]. This method combines the objectives of KM and SC in a quotient trace maximization form and solves the problem by eigen-decomposition. KL is shown to empirically outperforming KM and SC on real data sets. This straightforward idea is useful to solve many practical problems, especially those pertaining to combine attribute-based data with interaction-based networks. For example, in web analysis and scientometrics, the combination of text mining and bibliometrics has become a standard approach in clustering science or technology fields towards the detection of emerging fields or hot topics ([16]). In bioinformatics, protein-protein interaction network and expression data are two of the most important sources used to reveal the

S. Yu et al.: Kernel-based Data Fusion for Machine Learning, SCI 345, pp. 145–172.
springerlink.com 

relevance of genes and proteins with complex diseases. Conventionally, the data is often transformed into similarity matrices or interaction graphs, then consequently clustered by KM or SC. In KL, the similarity based kernel matrix and the interaction based Laplacian matrix are combined, which provides a novel approach to combine heterogeneous data structures in clustering analysis.

Our preliminary experiments show that when using KL to combine a single kernel and a single Laplacian, its performance strongly depends on the quality of the kernel and the Laplacian, which results in a model selection problem to determine the optimal settings of the kernel and the Laplacian. To perform model selection on unlabeled data is non-trivial because it is difficult to evaluate the models. To tackle the new problem, we propose a novel algorithm to incorporate multiple kernels and Laplacians in KL clustering. In chapter 4, we propose a method to integrate multiple kernel matrices in kernel $k$-means clustering. The main idea of this chapter lies in the additive combination of multiple kernels and Laplacians, moreover, the coefficients assigned to the kernels and the Laplacians are optimized automatically. This chapter presents the mathematical derivations of the additive integration form of kernels and Laplacians. The optimization of coefficients and clustering are achieved via a solution based on bi-level alternating minimization [5]. We validate the proposed algorithm on heterogeneous data sets taken from two real applications, where the advantage and reliability of the proposed method are systematically compared and demonstrated.

## 6.2 Acronyms

The symbols and notations used in this Chapter are defined in Table 6.1 to Table 6.4 (in the order of appearance).

**Table 6.1** Matrices

| | | | |
|---|---|---|---|
| $X$ | $\in$ | $\mathbb{R}^{N\times\mathscr{D}}$ | Data matrix with zero sample mean |
| $H$ | $\in$ | $\mathbb{R}^{N\times N}$ | Graph |
| $A$ | $\in$ | $\mathbb{R}^{N\times k}$ | Weighted scalar cluster membership matrix |
| $I_k$ | $\in$ | $\mathbb{R}^{k\times k}$ | Identity matrix |
| $I_N$ | $\in$ | $\mathbb{R}^{N\times N}$ | Identity matrix |
| $W$ | $\in$ | $\mathbb{R}^{N\times N}$ | Weighted adjacency matrix |
| $D$ | $\in$ | $\mathbb{R}^{N\times N}$ | Diagonal matrix whose $(a,a)$ entry is the sum of the entries of row $a$ in $W$ |
| $\tilde{L}$ | $\in$ | $\mathbb{R}^{N\times N}$ | Normalized Laplacian Matrix |
| $\hat{L}$ | $\in$ | $\mathbb{R}^{N\times N}$ | Normalized Laplacian Matrix |
| $G$ | $\in$ | $\mathbb{R}^{N\times N}$ | Kernel Matrix |
| $G_c$ | $\in$ | $\mathbb{R}^{N\times N}$ | Centered kernel Matrix |
| $P$ | $\in$ | $\mathbb{R}^{N\times N}$ | Centering Matrix |
| $\hat{\mathbf{L}}$ | $\in$ | $\mathbb{R}^{N\times N}$ | The combined Laplacian matrix of multiple $\hat{L}$ |
| $\mathbf{G}$ | $\in$ | $\mathbb{R}^{N\times N}$ | The combined matrix of multiple $G$ |
| $\Omega$ | $\in$ | $\mathbb{R}^{N\times N}$ | The combined matrix of kernels and Laplacians |
| $W$ | $\in$ | $\mathbb{R}^{N\times k}$ | The projection matrix determining the directions of discriminant hyperplanes |
| $F$ | $\in$ | $\mathbb{R}^{N\times k}$ | An affinity function matrix using +1 and -1 to discriminant pairwise clustering assignment |
| $Y$ | $\in$ | $\mathbb{R}^{N\times N}$ | Diagonal matrix where the $(a,a)$ entry using -1 or +1 to represent the cluster label of the $a-$th sample |
| $\mathbf{S_b}$ | $\in$ | $\mathbb{R}^{k\times k}$ | The between cluster scatter matrix |
| $\mathbf{S_t}$ | $\in$ | $\mathbb{R}^{k\times k}$ | The total scatter matrix |
| $\mathbf{S_b^{\phi}}$ | $\in$ | $\mathbb{R}^{k\times k}$ | The between cluster scatter matrix in $\mathscr{F}$ |
| $\mathbf{S_t^{\phi}}$ | $\in$ | $\mathbb{R}^{k\times k}$ | The total scatter matrix in $\mathscr{F}$ |

**Table 6.2** Vectors

| | | | |
|---|---|---|---|
| $\theta$ | $\in$ | $\mathbb{R}^{r+s}$ | Coefficient vector of Laplacians and kernels |
| $\mu$ | $\in$ | $\mathbb{R}^{\mathscr{D}}$ | The global sample mean vector of $X$ |
| $\mu^{\Phi}$ | $\in$ | $\mathbb{R}^{\mathscr{F}}$ | The global sample mean vector of $X^{\Phi}$ |
| $\mathbf{w}$ | $\in$ | $\mathbb{R}^{N}$ | Norm vector of the separating hyperplane in discriminant analysis |
| $\beta$ | $\in$ | $\mathbb{R}^{N}$ | Dual variables in convex optimization problems |
| $\mathbf{1}$ | $\in$ | $\mathbb{R}^{N}$ | Vector of all ones |
| $\mathbf{s}$ | $\in$ | $\mathbb{R}^{r}$ or $\mathbb{R}^{s}$ | Dummy vector in optimization problem |

**Table 6.3** Scalars

| | | | |
|---|---|---|---|
| $N$ | $\in$ | $\mathbb{N}$ | Number of data samples |
| $k$ | $\in$ | $\mathbb{N}$ | Number of clusters in *k*-means clustering |
| $a$ | $\in$ | $\{1,...,N\}$ | Index of the data samples |
| $b$ | $\in$ | $\{1,...,k\}$ | Index of the clusters |
| $i$ | $\in$ | $\{1,...,r\}$ | Index of the Graphs or Laplacians |
| $j$ | $\in$ | $\{1,...,s\}$ | Index of the kernels |
| $r$ | $\in$ | $\mathbb{N}$ | Number of Laplacians |
| $s$ | $\in$ | $\mathbb{N}$ | Number of kernels |
| $l$ | $\in$ | $\{1,...,r+s\}$ | Index of all sources (kernels and Laplacians) |
| $n_b$ | $\in$ | $\mathbb{N}$ | Number of samples belonging to cluster $b$ |
| $\kappa$ | $\in$ | $[0,1]$ | The parameter adjusting the effect of *k*-means and Spectral Clustering in the objective function |
| $\delta$ | $\in$ | $\{1,2\}$ | Sparseness control parameter |
| $t$ | $\in$ | $\mathbb{R}$ | Dummy variable in optimization problem |
| $\lambda$ | $\in$ | $\mathbb{R}^+$ | Regularization parameter in LSSVM |
| $\gamma$ | $\in$ | $\mathbb{Z}^+$ | Iteration index of OKLC algorithm |
| $\Delta A$ | $\in$ | $[0,1]$ | Error of clustering assignment matrix |
| $\varepsilon$ | $\in$ | $\mathbb{R}^+$ | Stopping criterion for OKLC |
| $\rho$ | $\in$ | $\mathbb{R}^+$ | Regularization parameter |

**Table 6.4** Others

| | |
|---|---|
| $\phi(\cdot)$ | A feature map |
| $\mathscr{K}(\cdot,\cdot)$ | kernel trick |
| $C$ | A set of samples in a cluster |
| $\mathscr{F}$ | Feature space |

## 6.3 Combine Kernel and Laplacian for Clustering

### 6.3.1 *Combine Kernel and Laplacian as Generalized Rayleigh Quotient for Clustering*

We first briefly review the KL algorithm proposed by [25]. Let us denote $X$ as an attribute data set and $W$ as a graph affinity matrix, both of them are representations of the same sets of samples. The objective of the KL integration to combine $X$ and $W$ for clustering can be defined as

$$J_{KL} = \kappa J_{SC} + (1-\kappa)J_{KM}, \tag{6.1}$$

where $J_{SC}$ and $J_{KM}$ are respectively the objectives of SC and KM clustering, $\kappa \in [0,1]$ is a coefficient adjusting the effect of the two objectives. Let us denote $A \in \mathbb{R}^{N\times k}$ as the weighted scalar cluster membership matrix, given by

$$A_{ab} = \begin{cases} \frac{1}{\sqrt{n_b}} & \text{if } \mathbf{x}_a \in C_b \\ 0 & \text{if } \mathbf{x}_a \notin C_b, \end{cases} \tag{6.2}$$

where $n_b$ is the number of data points belonging to cluster $C_b$ and $A^TA = I_k$, where $I_k$ denotes a $k\times k$ identity matrix. Let's denote $D$ as the diagonal matrix whose $(a,a)$ entry is the sum of the entries of row $a$ in the affinity matrix W. The *normalized Laplacian matrix* [31] is given by

$$\tilde{L} = I - D^{-\frac{1}{2}}WD^{-\frac{1}{2}}. \tag{6.3}$$

The objective of normalized cut based SC is formulated as

$$\underset{A}{\text{minimize}} \ \text{trace}\left(A^T\tilde{L}A\right). \tag{6.4}$$

As discussed in the literature [4, 8, 12], if the data $X$ has zero sample means, the objective of the KM is given by

$$\underset{A}{\text{maximize}} \ \text{trace}(A^TX^TXA). \tag{6.5}$$

We further generalize (6.5) by applying the feature map $\phi(\cdot):\mathbb{R}\rightarrow\mathscr{F}$ on $X$, then the centered data in $\mathscr{F}$ is denoted as $X^{\Phi}$, given by

$$X^{\Phi} = [\phi(\mathbf{x}_1)-\mu^{\Phi}, \phi(\mathbf{x}_2)-\mu^{\Phi}, ..., \phi(\mathbf{x}_N)-\mu^{\Phi}], \tag{6.6}$$

where $\phi(\mathbf{x}_i)$ is the feature map applied on the column vector of the $a$-th data point in $\mathscr{F}$, $\mu^{\Phi}$ is the global mean in $\mathscr{F}$. The inner product $X^TX$ in (6.5) can be combined using the kernel trick $\mathscr{K}(\mathbf{x}_u,\mathbf{x}_v) = \phi(\mathbf{x}_u)^T\phi(\mathbf{x}_v)$ and $G(\cdot,\cdot)$ is a Mercer kernel constructed by the kernel trick. We denote $G_c$ as the centered kernel matrix as $G_c = PGP$, where $P$ is the centering matrix $P = I_N - (1/N)\mathbf{1}_N^T$, $G$ is the kernel matrix, $I_N$ is the $N \times N$ identity matrix, $\mathbf{1}_N$ is a column vector of $N$ ones. Without loss of generality, the $k$-means objective in (6.5) can be equivalently written as

$$\underset{A}{\text{maximize}} \;\; \text{trace}(A^T G_c A). \tag{6.7}$$

Then the objective of KL integration becomes

$$\begin{aligned} \underset{A}{\text{minimize}} \;\; & \text{trace}\left(A^T \tilde{L} A\right) - (1-\kappa)\, \text{trace}\left(A^T G_c A\right) \\ \text{subject to} \;\; & A^T A = I_k, \\ & 0 \leq \kappa \leq 1. \end{aligned} \tag{6.8}$$

To solve the optimization problem without tuning the ad hoc parameter $\kappa$, Wang *et al.* formulate it as a trace quotient of the two components [25]. The trace quotient is then further relaxed as a maximization of quotient trace, given by

$$\begin{aligned} \underset{A}{\text{maximize}} \;\; & \text{trace} \;\; (A^T \tilde{L} A)^{-1} (A^T G_c A) \\ \text{subject to} \;\; & A^T A = I_k. \end{aligned} \tag{6.9}$$

The problem in (6.9) is a generalized Rayleigh quotient and the optimal solution $A^*$ is obtained in the generalized eigenvalue problem. To maximize this objective, $A^*$ is approximated as the largest $k$ eigenvectors of $\tilde{L}^+ G_c$, where $\tilde{L}^+$ is the pseudo inverse of $\tilde{L}$ [25].

### 6.3.2 Combine Kernel and Laplacian as Additive Models for Clustering

As discussed, the original KL algorithm is proposed to optimize the generalized Rayleigh quotient objective. In this chapter, we propose an alternative integration method using a different notation of Laplacian [31], $\hat{L}$, given by

$$\hat{L} = D^{-1/2} W D^{-1/2}, \tag{6.10}$$

where $D$ and $W$ are defined the same as in (6.3). The objective of spectral clustering is equivalent to maximizing the term as

$$\underset{A}{\text{maximize}} \;\; \text{trace}(A^T \hat{L} A). \tag{6.11}$$

Therefore, the objective of the KL integration can be rewritten in an additive form, given by

$$\begin{aligned} \underset{A}{\text{maximize}} \;\; & \text{trace}\left\{\kappa A^T \hat{L} A + (1-\kappa) A^T G_c A\right\} \\ \text{subject to} \;\; & A^T A = I_k, \\ & 0 \leq \kappa \leq 1, \end{aligned} \tag{6.12}$$

where $A$, $G_c$ are defined the same as in (6.8), $\kappa$ is the free parameter to adjust the effect of kernel and Laplacian in KL integration. If $\kappa$ is pre-defined, (6.12) is a Rayleigh quotient problem and the optimal $A^*$ can be obtained from eigenvalue decomposition, known as the spectral relaxation [7]. Therefore, to maximize this objective, we denote $\Omega = \kappa\hat{L} + (1-\kappa)G_c$ thus $A^*$ is solved as the dominant $k$ eigenvectors of $\Omega$.

We have shown two different methods to integrate a single Laplacian matrix with a single kernel matrix for clustering, where the main difference is to either optimize the cluster assignment affinity matrix $A$ as a generalized Rayleigh quotient (ratio model) or as a Rayleigh quotient (additive model). The main advantage of the ratio based solution is to avoid tuning the parameter $\kappa$. However, since our main interest is to optimize the combination of multiple kernels and Laplacians, the coefficients assigned on each kernel and Laplacian still need to be optimized. Moreover, the optimization of the additive integration model is computationally simpler than optimizing the ratio based model. Therefore, in the following sections we will focus on extending the additive KL integration to multiple sources.

## 6.4 Clustering by Multiple Kernels and Laplacians

Let us denote a set of graphs as $H_i$, $i \in \{1,...,r\}$, all having $N$ vertices, and a set of Laplacians $\hat{L}_i$ constructed from $H_i$ as (6.10). Let us also denote a set of centered kernel matrices as $G_{cj}$, $j \in \{1,...,s\}$ with $N$ samples. To extend (6.12) by incorporating multiple kernels and Laplacians for clustering, we propose a strategy to learn their optimal weighted convex linear combinations. The extended objective function is then given by

$$\boxed{\text{Q1:}} \quad \underset{A,\theta}{\text{maximize}} = \text{trace}\left(A^T(\hat{\mathbf{L}} + \mathbf{G})A\right) \tag{6.13}$$
$$\begin{aligned} \text{subject to } \; & \tilde{\mathbf{L}} = \sum_{i=1}^{r} \theta_i \tilde{L}_i, \\ & \mathbf{G} = \sum_{j=1}^{s} \theta_{j+r} G_{cj}, \\ & \sum_{i=1}^{r} \theta_i^\delta = 1, \; \sum_{j=1}^{s} \theta_{j+r}^\delta = 1, \\ & \theta_l \geq 0, \; l = 1,...,(r+s), \\ & A^T A = I_k, \end{aligned}$$

where $\theta_1,...,\theta_r$ and $\theta_{r+1},...,\theta_{r+s}$ are respectively the optimal coefficients assigned to the Laplacians and the kernels. $\mathbf{G}$ and $\tilde{\mathbf{L}}$ are respectively the combined kernel matrix and the combined Laplacian matrix. The $\kappa$ parameter in (6.12) is replaced by the coefficients assigned on each individual data sources.

To solve Q1, in the first phase we maximize $\mathscr{J}_{Q1}$ with respect to $A$, keeping $\theta$ fixed (initialized by random guess). In the second phase we maximize $\mathscr{J}_{Q1}$ with respect to $\theta$, keeping $A$ fixed. The two phases optimize the same objective and repeat until convergence locally. When $\theta$ is fixed, denoting $\Omega = \tilde{\mathbf{Ł}} + \tilde{\mathbf{G}}$, $Q_1$ is exactly a Rayleigh quotient problem and the optimal $A^*$ can solved as a eigenvalue problem of $\Omega$. When $A$ is fixed, the problem reduces to the optimization of the coefficients $\theta_l$ with given cluster memberships. In Chapter 4, we have shown that when the $A$ is given, Q1 can be formulated as Kernel Fisher Discriminant (KFD) in the high dimensional feature space $\mathscr{F}$. We introduce $W = [\mathbf{w}_1, ..., \mathbf{w}_k]$, a projection matrix determining the pairwise discriminating hyperplane. Since the discriminant analysis is invariant to the magnitude of $\mathbf{w}$, we assume that $W^T W = I_k$, thus Q1 can be equivalently formulated as

$$\boxed{\text{Q2:}}\ \underset{A,W,\theta}{\text{maximize}}\ \text{trace}\left(W^T A^T A W\right)^{-1}\left(W^T A^T (\mathbf{G} + \hat{\mathbf{Ł}}) A W\right), \tag{6.14}$$

$$\begin{aligned}
\text{subject to}\ & A^T A = I_k, \\
& W^T W = I_k, \\
& \hat{\mathbf{Ł}} = \sum_{i=1}^{r} \theta_i \hat{L}_i, \\
& \mathbf{G} = \sum_{j=1}^{s} \theta_{j+r} G_{cj}, \\
& \theta_l \geq 0,\ l = 1, ..., (r+s), \\
& \sum_{i=1}^{r} \theta_i^{\delta} = 1,\ \sum_{j=1}^{s} \theta_{j+r}^{\delta} = 1.
\end{aligned}$$

The bi-level optimization to solve Q1 correspond to two steps to solve Q2. In the first step (clustering), we set $W = I_k$ and optimize $A$, which is exactly the additive kernel Laplacian integration as (6.12); In the second step (KFD), we fix $A$ and optimize $W$ and $\theta$. Therefore, the two components optimize towards the same objective as a Rayleigh quotient in $\mathscr{F}$ so the iterative optimization converges to a local optimum. Moreover, in the second step, we are not interested in the separating hyperplane defined in $W$, instead, we only need the optimal coefficients $\theta_l$ assigned on the Laplacians and the kernels. In Chapter 4, we have known that Fisher Discriminant Analysis is related to the least squares approach [8], and the *Kernel Fisher discriminant* (KFD) ([17]) is related to and can be solved as a least squares support vector machine (LS-SVM), proposed by Suykens *et al.* [23]. The problem of optimizing multiple kernels for supervised learning (MKL) has been studied by [2, 15]. In our recent work [30], we derive the MKL extension for LSSVM and propose some efficient solutions to solve the problem. In this paper, the KFD problems are formulated as LSSVM MKL and solved by Semi-infinite programming (SIP) [21]. The concrete solutions and algorithms are presented in Chapter 3.

### 6.4.1 Optimize A with Given θ

When $\theta$ are given, the kernel-Laplacian combined matrix $\Omega$ is also fixed, therefore, the optimal $A$ can be found as the dominant $k$ number of eigenvectors of $\Omega$.

### 6.4.2 Optimize θ with Given A

When $A$ is given, the optimal $\theta$ assigned on Laplacians can be solved via the following KFD problem:

$$\boxed{\text{Q3:}}\quad \underset{W,\theta}{\text{maximize}}\ \text{trace}\left(W^T A^T A W\right)^{-1}\left(W^T A^T \hat{\mathfrak{L}} A W\right) \tag{6.15}$$
$$\text{subject to}\quad W^T W = I_k,$$
$$\hat{\mathfrak{L}} = \sum_{i=1}^{r} \theta_i \hat{L}_i,$$
$$\theta_i \geq 0,\ i = 1, ..., r,$$
$$\sum_{i=1}^{r} \theta_i^{\delta} = 1.$$

In Chapter 3, we have found that the $\delta$ parameter controls the sparseness of source coefficients $\theta_1, ..., \theta_r$. The issue of sparseness in MKL is also addressed by Kloft *et al.* ([14]) When $\delta$ is set to 1, the optimized solution is sparse, which assigns dominant values to only one or two Laplacians (kernels) and zero values to the others. The sparseness is useful to distinguish relevant sources from a large number of irrelevant data sources. However, in many applications, there are usually a small number of sources and most of these data sources are carefully selected and preprocessed. They thus often are directly relevant to the problem. In these cases, a sparse solution may be too selective to thoroughly combine the complementary information in the data sources. We may thus expect a non-sparse integration method which smoothly distributes the coefficients on multiple kernels and Laplacians and, at the same time, leverages their effects in the objective optimization. We have proved that when $\delta$ is set to 2, the KFD step in (6.15) optimizes the $L_2$-norm of multiple kernels, which yields a non-sparse solution. If we set $\delta$ to 0, the cluster objective is simplified as to averagely combine multiple kernels and Laplacians. In this chapter, we set $\delta$ to three different vales (0,1,2) to respectively optimize the sparse, average, and non-sparse coefficients on kernels and Laplacians.

When $\delta$ is set to 1, the KFD problem in Q3 is solved as LSSVM MKL [30], given by

$$\boxed{\text{Q4:}}\quad \underset{\beta,t}{\text{minimize}}\ \frac{1}{2}t + \frac{1}{2\lambda}\sum_{b=1}^{k} \beta_b^T \beta_b - \sum_{b=1}^{k} \beta_b^T Y_b^{-1} \mathbf{1} \tag{6.16}$$

$$\text{subject to} \sum_{a=1}^{N} \beta_{ab} = 0, \quad b = 1, ..., k,$$

$$t \geq \sum_{b=1}^{k} \beta_b^T \hat{L}_i \beta_b, \quad i = 1, ..., r, \; b = 1, ..., k,$$

where $\beta$ is the vector of dual variables, $t$ is a dummy variable in optimization, $a$ is the index of data samples, $b$ is the cluster label index of the discriminating problem in KFD, $Y_b$ is the diagonal matrix representing the binary cluster assignment, the vector on the diagonal of $Y_b$ is equivalent to the $b$-th column of a affinity matrix $F_{ab}$ using $\{+1, -1\}$ to discriminate the cluster assignments, given by

$$F_{ab} = \begin{cases} +1 & \text{if } A_{ab} > 0, \;\; a = 1, ..., N, \; b = 1, ..., k \\ -1 & \text{if } A_{ab} = 0, \;\; a = 1, ..., N, \; b = 1, ..., k \;. \end{cases} \tag{6.17}$$

As mentioned in Chapter 3, the problem presented in Q4 has an efficient solution based on SIP formulation. The optimal coefficients $\theta_i$ correspond to the dual variables bounded by the quadratic constraint $t \geq \sum_{b=1}^{k} \beta_b^T \hat{L}_i \beta_b$ in (6.16). When $\delta$ is set to 2, the solution to Q3 is given by

$$\boxed{\text{Q5:}} \; \underset{\beta, t}{\text{minimize}} \; \frac{1}{2} t + \frac{1}{2\lambda} \sum_{j=1}^{k} \beta_b^T \beta_b - \sum_{b=1}^{k} \beta_b^T Y_b^{-1} \mathbf{1} \tag{6.18}$$

$$\text{s.t.} \sum_{a=1}^{N} \beta_{ab} = 0, \quad b = 1, ..., k,$$

$$t \geq ||\mathbf{s}||_2,$$

where $\mathbf{s} = \{\sum_{b=1}^{k} \beta_b^T \hat{L}_1 \beta_b, ..., \sum_{b=1}^{k} \beta_b^T \hat{L}_r \beta_b\}^T$, other variables are defined the same as (6.16). The main difference between Q4 and Q5 is that Q4 optimizes the $L_\infty$ norm of multiple kernels whereas Q5 optimizes the $L_2$ norm. The optimal coefficients solved by Q4 are more likely to be sparse, in contrast, the ones obtained by Q5 are nonsparse. The algorithm to solve Q4 and Q5 is concretely explained in Algorithm 3.6.1 in Chapter 3.

Analogously, the coefficients assigned on kernels can also be obtained in the similar formulation, given by

$$\boxed{\text{Q6:}} \; \underset{W, \theta}{\text{maximize}} \; \text{trace} \left(W^T A^T A W\right)^{-1} \left(W^T A^T \mathbf{G} A W\right) \tag{6.19}$$

$$\text{subject to } W^T W = I_k,$$

$$\mathbf{G} = \sum_{j=1}^{s} \theta_{j+r} G_{cj},$$

$$\theta_{j+r} \geq 0, \; j = 1, ..., s,$$

$$\sum_{j=1}^{s} \theta_{j+r}^{\delta} = 1,$$

where most of the variables are defined in the similar way as Q3 in (6.15). The main difference is that the Laplacian matrices $\hat{\mathbf{L}}$ and $\hat{L}_i$ are replaced by the centered kernel matrices $\mathbf{G}$ and $G_{cj}$. The solution of Q6 is exactly the same as Q3, depending on the $\delta$ value, it can be solved either as Q4 or Q5.

### 6.4.3 Algorithm: Optimized Kernel Laplacian Clustering

As discussed, the proposed algorithm optimizes $A$ and $\theta$ iteratively to convergence. The coefficients assigned to the Laplacians and the kernels are optimized in parallel. Putting all the steps together, the pseudocode of the proposed *optimized kernel Laplacian clustering* (OKLC) is presented in Algorithm 6.4.1.

The iterations in Algorithm 6.4.1 terminate when the cluster membership matrix $A$ stops changing. The tolerance value $\varepsilon$ is a constant value as the stopping rule of OKLC and in our implementation it is set to 0.05. In our implementation, the final cluster assignment is obtained using $k$-means algorithm on $A^{(\gamma)}$. In Algorithm 6.4.1, we consider the $\delta$ as predefined values. When $\delta$ is set to 1 or 2, the SIP-LSSVM-MKL function optimizes the coefficients as the formulation in (6.16) or (6.18) respectively. It is also possible to combine Laplacians and kernels in an average manner. In this paper, we compare all these approaches and implement three different OKLC models. These three models are denoted as OKLC model 1, OKLC model 2, and OKLC model 3 which respectively correspond to the objective Q2 in (6.14) when $\delta = 1$, average combination, $\delta = 2$.

---

**Algorithm 6.4.1.** OKLC$(G_{c1},...,G_{cs},\hat{L}_1,...,\hat{L}_r,k)$

---

**comment:** Obtain the $\Omega^{(0)}$ using the initial guess of $\theta_1^{(0)},...,\theta_{r+s}^{(0)}$

$A^{(0)} \leftarrow$ EIGENVALUE DECOMPOSITION$(\Omega^{(0)},k)$

$\gamma = 0$

**while** $(\Delta A > \varepsilon)$

**do**
$$\begin{cases} step1: F^{(\gamma)} \leftarrow A^{(\gamma)} \\ step2: \theta_1^{(\gamma)},...,\theta_r^{(\gamma)} \leftarrow \text{SIP-LSSVM-MKL}(\hat{L}_1,...,\hat{L}_r,F^{(\gamma)}) \\ step3: \theta_{r+1}^{(\gamma)},...,\theta_{r+s}^{(\gamma)} \leftarrow \text{SIP-LSSVM-MKL}(G_{c1},...,G_{cs},F^{(\gamma)}) \\ step4: \Omega^{(r+1)} \leftarrow \theta_1^{(\gamma)}\hat{L}_1^{(\gamma)} + ... + \theta_r^{(\gamma)}\hat{L}_r^{(\gamma)} + \theta_{r+1}^{(\gamma)}G_{c1}^{(\gamma)} + ... + \theta_{r+s}^{(\gamma)}G_{cs}^{(\gamma)} \\ step5: A^{(\gamma+1)} \leftarrow \text{EIGENVALUE DECOMPOSITION}(\Omega^{(\gamma+1)},k) \\ step6: \Delta A = ||A^{(\gamma+1)} - A^{(\gamma)}||^2/||A^{(\gamma+1)}||^2 \\ step7: \gamma := \gamma + 1 \end{cases}$$

**return** $(A^{(\gamma)},\theta_1^{(\gamma)},...,\theta_r^{(\gamma)},\theta_{r+1}^{(\gamma)},...,\theta_{r+s}^{(\gamma)})$

---

## 6.5 Data Sets and Experimental Setup

The proposed OKLC models are validated in two real applications to combine heterogeneous data sets in clustering analysis. The data sets in the first experiment is taken from the work of multi-view text mining for disease gene identification mentioned in Chapter 5. The data sets contain nine different gene-by-term text profiles indexed by nine controlled vocabularies. The original disease relevant gene data set contains 620 genes which are known to be relevant to 29 diseases. To avoid the effect of imbalanced clusters which may affect the evaluation, we only keep the diseases who have 11 to 40 relevant genes (presented in Table 6.5). This results in 14 genetic diseases and 278 genes. Because the present paper is focused on non-overlapping "hard" clustering, we further remove 16 genes which are relevant to multiple diseases. The remaining 262 disease relevant genes are clustered into 14 clusters and evaluated biologically by their disease labels. For each vocabulary based gene-by-term data source, we create a kernel matrix using the linear kernel function and the *kernel normalization* method proposed by (Chapter 5, [19]). An element in the kernel matrix is then equivalent to the value of cosine similarity of two vectors [3]. This kernel is then regarded as the weighted adjacency matrix to create the Laplacian matrix. In total, nine kernels and nine Laplacian matrices are combined in clustering.

**Table 6.5** 14 genetic diseases in disease data and the number of genes relevant to each disease. The numbers in parentheses are the removed overlapping genes in each disease.

| Number | Disease | Number of genes |
|---|---|---|
| 1 | breast cancer | 24 (4) |
| 2 | cardiomyopathy | 22 (5) |
| 3 | cataract | 20 (0) |
| 4 | charcot marie tooth disease | 14 (4) |
| 5 | colorectal cancer | 21 (4) |
| 6 | diabetes | 26 (1) |
| 7 | emolytic anemia | 13 (0) |
| 8 | epilepsy | 15 (1) |
| 9 | lymphoma | 31 (4) |
| 10 | mental retardation | 24 (1) |
| 11 | muscular dystrophy | 24 (5) |
| 12 | neuropathy | 18 (3) |
| 13 | obesity | 13 (1) |
| 14 | retinitis pigmentosa | 30 (0) |

The data sets in the second experiment are taken from Web of Science (WOS) database provided by Thomson Scientific [16]. The original WOS data contains more than six million papers published from 2002 to 2006 (*e.g.*,articles, letters, notes, reviews) provided by Thomson Scientific. Citations received by these papers have been determined for a variable citation window beginning with the publication year, up to 2006. An item-by-item procedure was used with special identification-keys made up of bibliographic data elements, which were extracted from the first-author names, journal title, publication year, volume and the first page. To resolve ambiguities, journals were checked for the name changes and the papers were checked for name changes and merged accordingly. Journals not covered in the entire period (from 2002 to 2006) have been omitted. Two criteria were applied to select journals for clustering: at first, only the journals with at least 50 publications from 2002 to 2006 were investigated, and others were removed from the data set; then only those journals with more than 30 citations from 2002 to 2006 were kept. With these selection criteria, we obtained 8305 journals as the data set. We referred the ESI (Essential Science Index) labels of these journals and selected 7 balanced categories (1424 journals) as the journal set data in this paper. Table 6.6 shows these 7 ESI categorizations and the number of relevant journals.

**Table 6.6** 7 journal categorizations in journal data

| Number | ESI labels | Number of journals |
|---|---|---|
| 1 | Agriculture Science | 183 |
| 2 | Computer Science | 242 |
| 3 | Environment Ecology | 217 |
| 4 | Materials Science | 258 |
| 5 | Molecular Biology and Genetics | 195 |
| 6 | Neuroscience and Behavior | 194 |
| 7 | Pharmacology and Toxicology | 135 |

The titles, abstracts and keywords of the journal publications are indexed by a Jakarta Lucene based text mining program using no controlled vocabulary. The index contains 9,473,061 terms and we cut the Zipf curve of the indexed terms at the head and the tail to remove the rare term, stopwords and common words. These words are known as usually irrelevant, also noisy for the clustering purpose. After the Zipf cut, 669,860 meaningful terms are used to represent the journal in a vector space model where the terms are attributes and the weights are calculated as four weighting schemes: TF-IDF, IDF, TF and binary. The paper-by-term vectors are then aggregated to journal-by-term vectors as the representations of the lexical data. Therefore, we have obtained 4 data sources as the lexical information of journals.

The citations among the journals were analyzed in four different aspects.

- **Cross-citation:** Two papers are defined as cross-citation if one cites another, where the value is equivalent to the frequency of citations. We ignore the direction of citations by symmetrizing the cross-citation matrix.

- **Binary cross-citation:** To avoid the large amount of cross-citation in the journals with a lot of publications, we use binary values (1 or 0) to represent whether there is (or no) citation between two journals, termed as binary cross-citation metric.
- **Co-citation:** Two papers from different journals are defined as co-citation if they are cited simultaneously by a paper from the third journal. The co-citation frequency of the two papers equals to the number of other papers that cite them simultaneously.
- **Bibliographic coupling:** Two papers from different journals are considered as bibliographic coupling when they cite the same paper from a third journal. The coupling frequency equals to the number of papers they simultaneously cite together. The citations among papers are also aggregated to journal level.

Thus, we obtain different information sources about the journals. Four of them are text mining based attribute data and we construct the corresponding kernel matrices using linear kernel function, denoted as TFIDF, IDF, TF, Binary kernels. Four citation sources represent graph-based relationships among journals and we construct corresponding Laplacians, denoted as Cross-citation, co-citation, bibliograph coupling and binary cross citation Laplacians. The lexical similarities are represented as normalized linear kernel matrices (using the same methods applied on the disease data) and the citation metrics are regarded as weighted adjacency matrices to create the Laplacians. Totally four kernels and four Laplacians are combined on journal data.

The data sets used in our experiments are provided with labels, therefore the clustering performance is evaluated as comparing the automatic partitions with the labels using Adjusted Rand Index (ARI) ([13]) and Normalized Mutual Information (NMI) ([22]). To evaluate the ARI and NMI performance, we set $k = 14$ on disease data and $k = 7$ on journal data. We also tune the OKLC model using different $k$ values.

## 6.6 Results

We implemented the proposed OKLC models to integrate multiple kernels and Laplacians on disease data and journal set data. To compare the performance, we also applied the six popular ensemble clustering methods mentioned in Chapter 5 to combine the partitions of individual kernels and Laplacians as a consolidate partition. They are CSPA [22], HGPA [22], MCLA [22], QMI [24], EACAL [9], and AdacVote [1]. As shown in Table 6.7 and Table 6.8, the performance of OKLC algorithms is better than all the compared methods and the improvement is significant. On disease data, the best performance is obtained by OKLC model 1, which uses sparse coefficients to combine 9 text mining kernels and 9 Laplacians to identify disease relevant clusters (ARI: 0.5859, NMI: 0.7451). On journal data, all three OKLC models perform comparably well. The best one seems coming from OKLC model 3 (ARI: 0.7336, NMI: 0.7758), which optimizes the non-sparse coefficients on the 4 kernels and 4 Laplacians.

**Table 6.7** Performance on disease data set. All the comparing methods combine nine kernels and nine Laplacians. The mean values and the standard deviations are observed from 20 random repetitions. The best performance is shown in bold. The P-values are statistically evaluated with the best performance using paired t-test.

| Algorithm | ARI | P-value | NMI | P-value |
|---|---|---|---|---|
| OKLC 1 | **0.5859**±0.0390 | - | **0.7451**±0.0194 | - |
| OKLC 2 | 0.5369±0.0493 | 2.97E-04 | 0.7106±0.0283 | 9.85E-05 |
| OKLC 3 | 0.5469±0.0485 | 1.10E-03 | 0.7268±0.0360 | 2.61E-02 |
| CSPA | 0.4367±0.0266 | 5.66E-11 | 0.6362±0.0222 | 4.23E-12 |
| HGPA | 0.5040±0.0363 | 8.47E-07 | 0.6872±0.0307 | 7.42E-07 |
| MCLA | 0.4731±0.0320 | 2.26E-10 | 0.6519±0.0210 | 5.26E-14 |
| QMI | 0.4656±0.0425 | 7.70E-11 | 0.6607±0.0255 | 8.49E-11 |
| EACAL | 0.4817±0.0263 | 2.50E-09 | 0.6686±0.0144 | 5.54E-12 |
| AdacVote | 0.1394±0.0649 | 1.47E-16 | 0.4093±0.0740 | 6.98E-14 |

**Table 6.8** Performance on journal data set. All the comparing methods combine four kernels and four Laplacians. The mean values and the standard deviations are observed from 20 random repetitions. The best performance is shown in bold. The P-values are statistically evaluated with the best performance using paired t-test.

| Algorithm | ARI | P-value | NMI | P-value |
|---|---|---|---|---|
| OKLC 1 | 0.7346±0.0584 | 0.3585 | 0.7688±0.0364 | 0.1472 |
| OKLC 2 | 0.7235±0.0660 | 0.0944 | 0.7532±0.0358 | 0.0794 |
| OKLC 3 | **0.7336**±0.0499 | - | **0.7758**±0.0362 | - |
| CSPA | 0.6703±0.0485 | 8.84E-05 | 0.7173±0.0291 | 1.25E-05 |
| HGPA | 0.6673±0.0419 | 4.74E-06 | 0.7141±0.0269 | 5.19E-06 |
| MCLA | 0.6571±0.0746 | 6.55E-05 | 0.7128±0.0463 | 2.31E-05 |
| QMI | 0.6592±0.0593 | 5.32E-06 | 0.7250±0.0326 | 1.30E-05 |
| EACAL | 0.5808±0.0178 | 3.85E-11 | 0.7003±0.0153 | 6.88E-09 |
| AdacVote | 0.5899±0.0556 | 1.02E-07 | 0.6785±0.0325 | 6.51E-09 |

To evaluate whether the combination of kernel and Laplacian indeed improve the clustering performance, we first systematically compared the performance of all the individual data sources using KM and SC. As shown in Table 6.9, on disease data, the best KM performance (ARI 0.5441, NMI 0.7099) and SC (ARI 0.5199, NMI 0.6858) performance is obtained on LDDB text mining profile. Next, we enumerated all the paired combinations of a single kernel and a single Laplacian for clustering. The integration is based on equation (6.12) and the $\kappa$ value is set to 0.5 so the objectives of KM and SC are combined averagely. The performance of all 45 paired combinations is presented in Table 6.10. As shown, the best KL clustering performance is obtained by integrating the LDDB kernel with KO Laplacian (ARI 0.5298, NMI 0.6949). Moreover, we also found that the integration performance varies significantly by the choice of kernel and Laplacian, which proves our previous point that the KL performance is highly dependent on the quality of kernel and Laplacian. Using the proposed OKLC algorithm, there is no need to enumerate all the possible paired combinations. OKLC combines all the kernels and Laplacians and optimizes their coefficients in parallel, yielding a comparable performance with the best paired combination of a single kernel and a single Laplacian.

**Table 6.9** Clustering performance of individual disease data source. The mean values and the standard deviations observed from 20 random repetitions are reported in the table. The KM clustering is implemented as the kernel $k$-means algorithm proposed by Girolami [10]. The SC is solved as equation (6.11). The bold numbers represent the best performance of kernels and Laplacians respectively.

| | Source | ARI | NMI |
|---|---|---|---|
| kernels | eVOC | 0.3880 ± 0.0469 | 0.6122 ± 0.0404 |
| | GO | 0.2801 ± 0.0282 | 0.4870 ± 0.0218 |
| | KO | 0.1233 ± 0.0186 | 0.3312 ± 0.0186 |
| | LDDB | **0.5441** ± 0.0374 | **0.7099** ± 0.0208 |
| | MeSH | 0.3288 ± 0.0471 | 0.5432 ± 0.0318 |
| | MP | 0.4234 ± 0.0481 | 0.6259 ± 0.0362 |
| | OMIM | 0.4185 ± 0.0454 | 0.6340 ± 0.0289 |
| | SNOMED | 0.3065 ± 0.0426 | 0.5357 ± 0.0340 |
| | Uniprot | 0.3684 ± 0.0352 | 0.5736 ± 0.0152 |
| Laplacians | eVOC | 0.4153 ± 0.0435 | 0.6106 ± 0.0312 |
| | GO | 0.3423 ± 0.0359 | 0.5440 ± 0.0266 |
| | KO | 0.1327 ± 0.0167 | 0.3323 ± 0.0169 |
| | LDDB | **0.5199** ± 0.0488 | **0.6858** ± 0.0269 |
| | MeSH | 0.4333 ± 0.0492 | 0.6403 ± 0.0360 |
| | MP | 0.4741 ± 0.0347 | 0.6561 ± 0.0237 |
| | OMIM | 0.4814 ± 0.0612 | 0.6806 ± 0.0312 |
| | SNOMED | 0.4240 ± 0.0420 | 0.6253 ± 0.0293 |
| | Uniprot | 0.3222 ± 0.0286 | 0.5374 ± 0.0234 |

**Table 6.10** KL performance of disease data. The mean values and the standard deviations observed from 20 random repetitions are reported in the table. The KL clustering is solved as equation (12) defined in the main manuscript. The $\kappa$ value is set to 0.5. The best performance among the 45 paired kernel-Laplacian integrations is represented in bold. Comparing the best KL result with the best KM and SC results in the previous table, it is obvious that KL integration does indeed improve the quality of clustering.

| ARI | L-eVOC | L-GO | L-KO | L-LDDB | L-MeSH | L-MP | L-OMIM | L-SNOMED | L-Uniprot |
|---|---|---|---|---|---|---|---|---|---|
| **K-eVOC** | 0.4073±0.04 | | | | | | | | |
| **K-GO** | 0.3347±0.04 | 0.3113±0.03 | | | | | | | |
| **K-KO** | 0.1366±0.01 | 0.1260±0.02 | 0.1193±0.01 | | | | | | |
| **K-LDDB** | 0.5630±0.04 | 0.5487±0.05 | 0.5386±0.06 | 0.5281±0.05 | | | | | |
| **K-MeSH** | 0.4375±0.04 | 0.4208±0.05 | 0.4270±0.04 | 0.5186±0.04 | 0.4247±0.04 | | | | |
| **K-MP** | 0.4718±0.08 | 0.4827±0.04 | 0.4649±0.04 | 0.5797±0.05 | 0.4801±0.05 | 0.4583±0.05 | | | |
| **K-OMIM** | 0.4615±0.05 | 0.4833±0.05 | 0.5065±0.05 | **0.5876**±0.05 | 0.4504±0.05 | 0.4645±0.05 | 0.4375±0.05 | | |
| **K-SNOMED** | 0.4143±0.03 | 0.4179±0.04 | 0.4007±0.03 | 0.5270±0.05 | 0.4022±0.03 | 0.4189±0.03 | 0.3933±0.04 | 0.4200±0.03 | |
| **K-Uniprot** | 0.3332±0.03 | 0.3236±0.03 | 0.3201±0.03 | 0.4083±0.04 | 0.3246±0.04 | 0.3161±0.03 | 0.3225±0.03 | 0.3254±0.02 | 0.3255±0.03 |

| NMI | L-eVOC | L-GO | L-KO | L-LDDB | L-MeSH | L-MP | L-OMIM | L-SNOMED | L-Uniprot |
|---|---|---|---|---|---|---|---|---|---|
| **K-eVOC** | 0.5988±0.03 | | | | | | | | |
| **K-GO** | 0.5513±0.03 | 0.5257±0.02 | | | | | | | |
| **K-KO** | 0.3368±0.01 | 0.3263±0.02 | 0.3223±0.01 | | | | | | |
| **K-LDDB** | 0.7310±0.03 | 0.7151±0.04 | 0.7200±0.03 | 0.6916±0.03 | | | | | |
| **K-MeSH** | 0.6380±0.03 | 0.6208±0.03 | 0.6357±0.03 | 0.7063±0.03 | 0.6276±0.02 | | | | |
| **K-MP** | 0.6529±0.05 | 0.6639±0.02 | 0.6547±0.02 | 0.7466±0.03 | 0.6594±0.03 | 0.6415±0.03 | | | |
| **K-OMIM** | 0.6685±0.03 | 0.6849±0.03 | 0.6996±0.03 | **0.7591**±0.02 | 0.6653±0.03 | 0.6734±0.03 | 0.6563±0.03 | | |
| **K-SNOMED** | 0.6255±0.03 | 0.6147±0.02 | 0.6067±0.02 | 0.6946±0.03 | 0.6104±0.03 | 0.6166±0.02 | 0.6058±0.03 | 0.6208±0.03 | |
| **K-Uniprot** | 0.5525±0.02 | 0.5437±0.03 | 0.5355±0.03 | 0.6057±0.03 | 0.5419±0.02 | 0.5433±0.02 | 0.5411±0.02 | 0.5489±0.01 | 0.5407±0.02 |

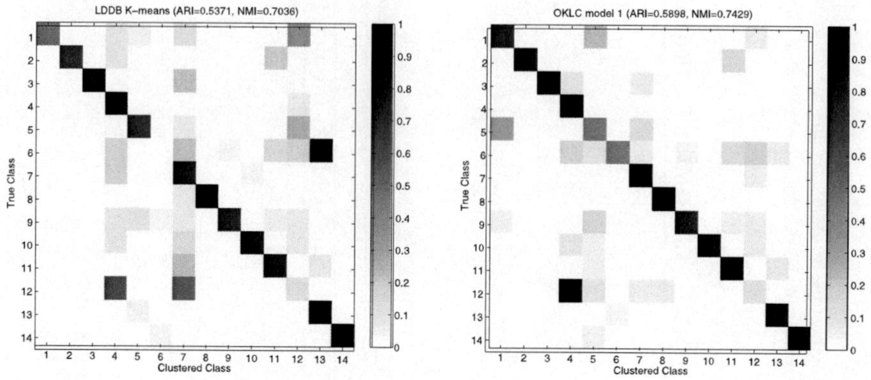


**Fig. 6.1** Confusion matrices of disease data obtained by kernel $k$-means on LDDB (figure on the left) and OKLC model 1 integration (figure on the right). The numbers of cluster labels are consistent with the numbers of diseases presented in Table 6.5. In each row of the confusion matrix, the diagonal element represents the fraction of correctly clustered genes and the off-diagonal non-zero element represents the fraction of mis-clustered genes.

In Figure 6.1 two confusion matrices of disease data for a single run are depicted. The values on the matrices are normalized according to $R_{ij} = C_j/T_i$, where $T_i$ is the total number of genes belonging in disease $i$ and $C_j$ is the number of these $T_i$ genes that were clustered to belong to class $j$. First, it is worth noting that OKLC reduces the number of mis-clustered genes on breast cancer (Nr.1), cardiomyopathy (Nr.2), and muscular dystrophy (Nr.11). Among the mis-clustered genes in LDDB, five genes (TSG101, DBC1, CTTN, SLC22A18, AR) in breast cancer, two genes in cardiomyopathy (COX15, CSRP3), and two genes in muscular dystrophy (SEPN1, COL6A3) are correctly clustered in OKLC model 1. Second, there are several diseases where consistent misclustering occurs in both methods, such as diabetes (Nr.6) and neuropathy (Nr.12). The intuitive confusion matrices correspond to the numerical evaluation results, as shown, the quality of clustering obtained by OKLC model 1 (ARI=0.5898, NMI=0.7429) is higher than LDDB, the best individual data.

The performance of individual data sources of journal data is shown in Table 6.11. The best KM (ARI 0.6482, NMI 0.7104) is obtained on the IDF kernel and the best SC (ARI 0.5667, NMI 0.6807) is obtained on the cross-citation Laplacian. To combine the 4 kernels with 4 Laplacians, we evaluate all the 10 paired combinations and show the performance in Table 6.12. The best performance is obtained by integrating the IDF kernel with the cross-citation Laplacian (ARI 0.7566, NMI 0.7702). As shown, the integration of lexical similarity information and citation based Laplacian indeed improves the performance.

In Figure 6.2 the confusion matrices (also normalized) of journal data for a single run are illustrated. We compare the best individual data source (IDF with kernel $k$-means, figure on the left) with the OKLC model 1. In the confusion matrix of

**Table 6.11** Clustering performance of individual journal data source. The mean values and the standard deviations observed from 20 random repetitions are reported in the table. The KM clustering is implemented as the kernel $k$-means algorithm proposed by Girolami [10]. The SC is solved as equation (6.11). The bold numbers represent the best performance of kernels and Laplacians respectively.

| | Source | ARI | NMI |
|---|---|---|---|
| Text kernels | IDF | **0.6482**±0.0506 | **0.7104**±0.0375 |
| | TFIDF | 0.5540±0.0760 | 0.6547±0.0446 |
| | TF | 0.5832±0.0643 | 0.6740±0.0413 |
| | Binary | 0.6458±0.0382 | 0.6949±0.0246 |
| Laplacians | Cross citation | **0.5667** ±0.0355 | **0.6870**±0.0192 |
| | Co-citation | 0.3239±0.0211 | 0.4941±0.0154 |
| | Bibliograph coupling | 0.5721±0.0295 | 0.6534±0.0162 |
| | Binary cross citation | 0.5037±0.0191 | 0.6221±0.0090 |

**Table 6.12** KL performance of journal data. The mean values and the standard deviations observed from 20 random repetitions are reported in the table. The KL clustering is solved as equation (6.12). The $\kappa$ value is set to 0.5. The best performance among the 10 paired kernel-Laplacian integrations is represented in bold. Comparing the best KL result with the best KM and SC results in the previous table, it is obvious that KL integration does indeed improve the quality of clustering.

| ARI | **L-crs-citation** | **L-co-citation** | **L-biblio** | **L-binary-crs-citation** |
|---|---|---|---|---|
| **K-IDF** | **0.7566**±0.09 | | | |
| **K-TFIDF** | 0.6830±0.04 | 0.7090±0.04 | | |
| **K-TF** | 0.7086±0.03 | 0.6819±0.05 | 0.6948±0.05 | |
| **K-Binary** | 0.6326±0.05 | 0.6597±0.05 | 0.6350±0.06 | 0.6494±0.07 |

| NMI | **L-crs-citation** | **L-co-citation** | **L-biblio** | **L-binary-crs-citation** |
|---|---|---|---|---|
| **K-IDF** | **0.7702**±0.07 | | | |
| **K-TFIDF** | 0.7347±0.03 | 0.7482±0.03 | | |
| **K-TF** | 0.7448±0.03 | 0.7237±0.03 | 0.7346±0.03 | |
| **K-Binary** | 0.6731±0.03 | 0.6852±0.03 | 0.6741±0.04 | 0.6802±0.04 |

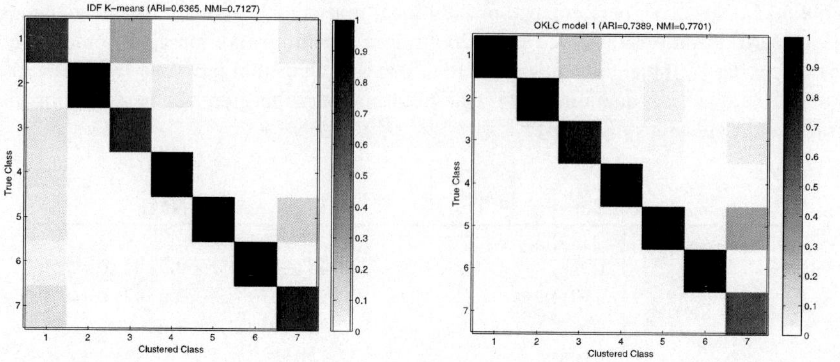


**Fig. 6.2** Confusion matrices of journal data obtained by kernel $k$-means on IDF (figure on the left) and OKLC model 1 integration (figure on the right). The numbers of cluster labels are consistent with the numbers of ESI journal categories presented in Table 6.6. In each row, the diagonal element represents the fraction of correctly clustered journals and the off-diagonal non-zero element represents the fraction of mis-clustered journals.

IDF $k$-means, 79 journals belonging to agriculture science (Nr.1) are mis-clustered to environment ecology (Nr.3), 9 journals are mis-clustered to pharmacology and toxicology (Nr.7). In OKLC, the number of agriculture journals mis-clustered to environment ecology is reduced to 45, and the number to pharmacology and toxicology is reduced to 5. On other journal clusters, the performance of the two models is almost equivalent.

We also investigated the performance of only combining multiple kernels or multiple Laplacians. On the disease data set, we respectively combined the 9 kernels and the 9 Laplacians for clustering, using all the compared methods in Table 6.7 and Table 6.8. On the journal data set, we combined the 4 text mining kernels and the 4 citation Laplacians. The proposed OKLC method is simplified as only optimizing coefficients on Laplacians (step 2 in Algorithm 6.4.1) or kernels (step 3). As shown in Table 6.13 and Table 6.14, the performance of OKLC is also comparable to the best performance obtained either by kernel combination or Laplacian combination. In particular, of all the methods we compared, the best performance is all obtained on OKLC models or its simplified forms.

**Table 6.13** Clustering results on disease data set. The table presents the clustering performance using a variety of methods. The performance of clustering by multiple kernels integration, multiple Laplacians integration and multiple kernels Laplacians integration is compared. The best performance of each approach is represented in Bold. As shown, two approaches lead to the best performance: One is the average combination of 9 Laplacian matrices (OKLC model 2). Another approach is the integration of multiple kernels and Laplacians using OKLC model 1.

| | Algorithm | ARI | NMI |
|---|---|---|---|
| kernels only | OKLC 1 | 0.4906±0.0356 | 0.6837±0.0267 |
| | OKLC 2 | **0.5362**±0.0546 | **0.7014**±0.0243 |
| | OKLC 3 | 0.4029±0.0353 | 0.6485±0.0194 |
| | CSPA | 0.4040±0.0197 | 0.6123±0.0204 |
| | HGPA | 0.4312±0.0451 | 0.6387±0.0401 |
| | MCLA | 0.4377±0.0331 | 0.6323±0.0223 |
| | QMI | 0.4309±0.0291 | 0.6328±0.0209 |
| | EACAL | 0.4563±0.0240 | 0.6606±0.0206 |
| | AdacVote | 0.2361±0.0805 | 0.5262±0.0702 |
| Laplacians only | OKLC 1 | 0.4603±0.0356 | 0.6881±0.0104 |
| | OKLC 2 | **0.6187**±0.0497 | **0.7665**±0.0226 |
| | OKLC 3 | 0.4857±0.0198 | 0.7137±0.0093 |
| | CSPA | 0.4348±0.0249 | 0.6368±0.0213 |
| | HGPA | 0.4323±0.0300 | 0.6326±0.0211 |
| | MCLA | 0.4916±0.0265 | 0.6609±0.0187 |
| | QMI | 0.4638±0.0416 | 0.6491±0.0248 |
| | EACAL | 0.5080±0.0217 | 0.6823±0.0167 |
| | AdacVote | 0.3933±0.1148 | 0.6019±0.1520 |
| kernels + Laplacians | OKLC 1 | **0.5859**±0.0390 | **0.7451**±0.0194 |
| | OKLC 2 | 0.5369±0.0493 | 0.7106±0.0283 |
| | OKLC 3 | 0.5469±0.0485 | 0.7268±0.0360 |
| | CSPA | 0.4367±0.0266 | 0.6362±0.0222 |
| | HGPA | 0.5040±0.0363 | 0.6872±0.0307 |
| | MCLA | 0.4731±0.0320 | 0.6519±0.0210 |
| | QMI | 0.4656±0.0425 | 0.6607±0.0255 |
| | EACAL | 0.4817±0.0263 | 0.6686±0.0144 |
| | AdacVote | 0.1394±0.0649 | 0.4093±0.0740 |

It is interesting to observe that the average combination model (OKLC model 2) performs quite well on the journal data set but not on the disease data set. This is probably because most of the sources in journal data set are relevant to the problem whereas in disease data set some data sources are noisy, thus the integration of disease data sources is a non-trivial task. We expect that the other two OKLC models (model 1 and model 3) optimize the coefficients assigned on the kernels and the Laplacians to leverage multiple sources in integration, at the same time, to increase

**Table 6.14** Clustering results on journal data set. The settings is the same as in Table 6.14.

| | Algorithm | ARI | NMI |
|---|---|---|---|
| kernels only | OKLC 1 | **0.7328**±0.0561 | **0.7756**±0.0408 |
| | OKLC 2 | 0.6428±0.0861 | 0.7014±0.0567 |
| | OKLC 3 | 0.6968±0.0953 | 0.7509±0.0531 |
| | CSPA | 0.6523±0.0475 | 0.7038±0.0283 |
| | HGPA | 0.6668±0.0621 | 0.7098±0.0334 |
| | MCLA | 0.6507±0.0639 | 0.7007±0.0343 |
| | QMI | 0.6363±0.0683 | 0.7058±0.0481 |
| | EACAL | 0.6670±0.0586 | 0.7231±0.0328 |
| | AdacVote | 0.6617±0.0542 | 0.7183±0.0340 |
| Laplacians only | OKLC 1 | 0.4390±0.0185 | 0.5818±0.0101 |
| | OKLC 2 | **0.7235**±0.0521 | **0.7630**±0.0280 |
| | OKLC 3 | 0.4934±0.0116 | 0.6351±0.0085 |
| | CSPA | 0.4797±0.0632 | 0.5852±0.0380 |
| | HGPA | 0.4674±0.0570 | 0.5772±0.0370 |
| | MCLA | 0.4870±0.0530 | 0.5865±0.0267 |
| | QMI | 0.5691±0.0430 | 0.6542±0.0301 |
| | EACAL | 0.5529±0.0265 | 0.6813±0.0122 |
| | AdacVote | 0.5197±0.0783 | 0.6537±0.0336 |
| kernels + Laplacians | OKLC 1 | **0.7346**±0.0584 | **0.7688**±0.0364 |
| | OKLC 2 | **0.7235**±0.0660 | **0.7532**±0.0358 |
| | OKLC 3 | **0.7336**±0.0499 | **0.7758**±0.0362 |
| | CSPA | 0.6703±0.0485 | 0.7173±0.0291 |
| | HGPA | 0.6673±0.0419 | 0.7141±0.0269 |
| | MCLA | 0.6571±0.0746 | 0.7128±0.0463 |
| | QMI | 0.6592±0.0593 | 0.7250±0.0326 |
| | EACAL | 0.5808±0.0178 | 0.7003±0.0153 |
| | AdacVote | 0.5899±0.0556 | 0.6785±0.0325 |

the robustness of the combined model on combining relevant and irrelevant data sources. To evaluate whether the optimized weights assigned on individual sources have correlation with the performance, we compare the rank of coefficients with the rank of performance from Table 6.15 to Table 6.18. As shown, the largest coefficients correctly indicate the best individual data sources. It is worth noting that in multiple kernel learning, the rank of coefficients are only moderately correlated with the rank of individual performance. In our experiments, the MeSH kernel gets the 2nd largest weights though its performance in evaluation is low. In MKL, it is usual that the best individual kernel found by cross-validation may not lead to a large weight when used in combination [27]. Kernel fusion combines multiple sources at a refined granularity, where the "moderate" kernels containing weak and

**Table 6.15** The average values of coefficients of kernels and Laplacians in disease data set optimized by OKLC model 1. The sources assigned with 0 coefficient are not presented. The performance is ranked by the average values of ARI and NMI evaluated on each individual sources presented in Table 6.9.

| Rank of $\theta$ | Source | $\theta$ value | Performance rank |
|---|---|---|---|
| 1 | LDDB kernel | 0.6113 | 1 |
| 2 | MESH kernel | 0.3742 | 6 |
| 3 | Uniprot kernel | 0.0095 | 5 |
| 4 | Omim kernel | 0.0050 | 2 |
| 1 | LDDB Laplacian | 1 | 1 |

**Table 6.16** The average values of coefficients of kernels and Laplacians in journal data set optimized by OKLC model 1. The sources assigned with 0 coefficient are not presented. The performance is ranked by the average values of ARI and NMI evaluated on each individual sources presented in Table 6.11.

| Rank of $\theta$ | Source | $\theta$ value | Performance rank |
|---|---|---|---|
| 1 | IDF kernel | 0.7574 | 1 |
| 2 | TF kernel | 0.2011 | 3 |
| 3 | Binary kernel | 0.0255 | 2 |
| 4 | TF-IDF kernel | 0.0025 | 4 |
| 1 | Bibliographic Laplacian | 1 | 1 |

insignificant information could complement to other kernels to compose a "good" kernel containing strong and significant information. Though such complementary information cannot be incorporated when cross-validation is used to choose a single best kernel, these "moderate" kernels are still useful when combined with other kernels [27]. Based on the ranks presented in Table 6.17 and Table 6.18, we calculated the spearman correlations between the ranks of weights and the ranks of performance on both data sets. The correlations of disease kernels, disease Laplacians, journal kernels and journal Laplacians are respectively 0.5657, 0.6, 0.8, and 0.4. In some relevant work, the average spearman correlations are mostly around 0.4 [15, 27]. Therefore, the optimal weights obtained in our experiments are generally consistent with the rank of performance.

**Table 6.17** The average values of coefficients of kernels and Laplacians in disease data set optimized by OKLC model 3

| Rank of $\theta$ | Source | $\theta$ value | Performance rank |
|---|---|---|---|
| 1 | LDDB kernel | 0.4578 | 1 |
| 2 | MESH kernel | 0.3495 | 6 |
| 3 | OMIM kernel | 0.3376 | 2 |
| 4 | SNOMED kernel | 0.3309 | 7 |
| 5 | MPO kernel | 0.3178 | 3 |
| 6 | GO kernel | 0.3175 | 8 |
| 7 | eVOC kernel | 0.3180 | 4 |
| 8 | Uniprot kernel | 0.3089 | 5 |
| 9 | KO kernel | 0.2143 | 9 |
| 1 | LDDB Laplacian | 0.6861 | 1 |
| 2 | MESH Laplacian | 0.2799 | 4 |
| 3 | OMIM Laplacian | 0.2680 | 2 |
| 4 | GO Laplacian | 0.2645 | 7 |
| 5 | eVOC Laplacian | 0.2615 | 6 |
| 6 | Uniprot Laplacian | 0.2572 | 8 |
| 7 | SNOMED Laplacian | 0.2559 | 5 |
| 8 | MPO Laplacian | 0.2476 | 3 |
| 9 | KO Laplacian | 0.2163 | 9 |

**Table 6.18** The average values of coefficients of kernels and Laplacians in journal data set optimized by OKLC model 3

| Rank of $\theta$ | Source | $\theta$ value | Performance rank |
|---|---|---|---|
| 1 | IDF kernel | 0.5389 | 1 |
| 2 | Binary kernel | 0.4520 | 2 |
| 3 | TF kernel | 0.2876 | 4 |
| 4 | TF-IDF kernel | 0.2376 | 3 |
| 1 | Bibliographic Laplacian | 0.7106 | 1 |
| 2 | Cocitation Laplacian | 0.5134 | 4 |
| 3 | Crosscitation Laplacian | 0.4450 | 2 |
| 4 | Binarycitation Laplacian | 0.1819 | 3 |

We also investigated the dominant eigenvalues of the optimized combination of kernels and Laplacians. In Figure 6.3, we compare the difference of three OKLC models with the pre-defined $k$ (set as equal to the number of class labels). In practical research, one can predict the optimal cluster number by checking the "elbow" of the eigenvalue plot. As shown, the "elbow" in disease data is quite obvious at the number of 14. In journal data, the "elbow" is more likely to range from 6 to 12. All the three OKLC models show a similar trend on the eigenvalue plot. Moreover, in Figure 6.4 and Figure 6.5, we also compare the eigenvalue curves using different $k$ values as input. As shown, the eigenvalue plot is quite stable w.r.t. the different inputs of $k$, which means the optimized kernel and Laplacian coefficients are quite independent with the $k$ value. This advantage enables a reliable prediction about the optimal cluster number by integrating multiple data sources.

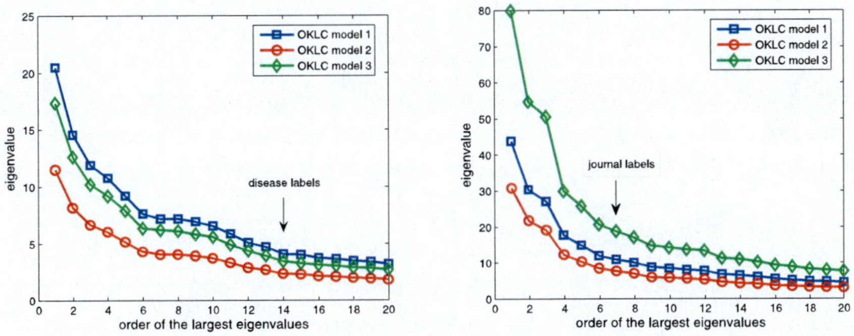


**Fig. 6.3** The plot of eigenvalues (disease data on the left, journal data on the right) of the optimal kernel-Laplacian combination obtained by all OKLC models. The parameter $k$ is set as equivalent as the reference label numbers.

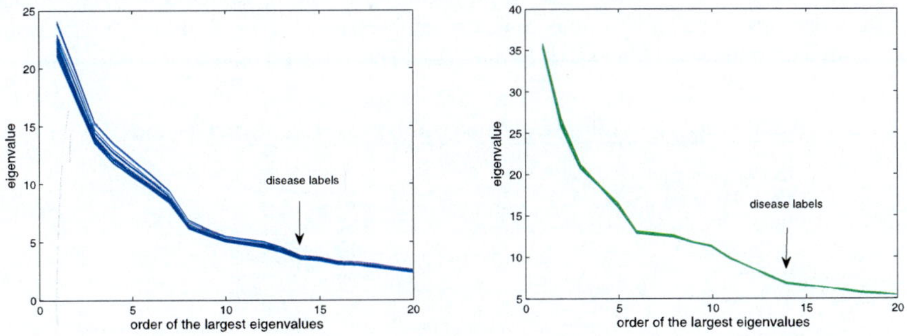


**Fig. 6.4** The plot of eigenvalues of the optimal kernel-Laplacian combination obtained by OKLC model 1 (left) and OKLC model 3 (right). We tried 19 different $k$ inputs from 2 to 20. For each $k$, the dominant 20 eigenvalues are shown as a curve. The figure on the top shows the curves of OKLC model 1. The figure on the bottom shows the curves of OKLC model 3. As shown, the eigenvlaue curves are insensitive to the input of $k$. No matter what $k$ value is used as the input, the resulted eigenvalue curves all show some obvious "elbows". It is also obvoious that the number of labeled diseases (14) is an "elbow".

To investigate the computational time, we benchmarked OKLC algorithms with other clustering methods on the two data sets. As shown in Table 6.19, when optimizing the coefficients, OKLC algorithm (model 1 and model 3) spends longer time than the other methods to optimize the coefficients on the Laplacians and the kernels. However, the proposed algorithm is still efficient. Considering the fact that the proposed algorithm yields much better performance and more enriched information (the ranking of the individual sources) than other methods, it is worth spending extra computational complexity on a promising algorithm.

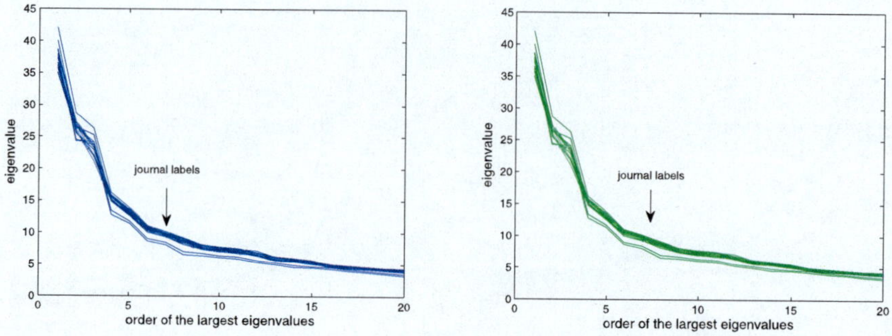


**Fig. 6.5** Similarly, this figure shows the eigenvalue curves obtained by OKLC model 1 (left) and OKLC model 3 (right) on journal data. Comparing with disease data, the optimal cluster number of journal data is not so obvious. If there is no ESI label information, one may predict the optimal cluster number from 6 to 8.

**Table 6.19** Comparison of CPU time of all algorithms. The reported values are averaged from 20 repetitions. The CPU time is evaluated on Matlab v7.6.0 + Windows XP2 installed on a Laptop computer with Intel Core 2 Duo 2.26G Hz and 2G memory.

| Algorithm | disease data (seconds) | journal data (seconds) |
|---|---|---|
| OKLC model 1 | 42.39 | 1011.4 |
| OKLC model 2 | 0.19 | 13.27 |
| OKLC model 3 | 37.74 | 577.51 |
| CSPA | 9.49 | 177.22 |
| HGPA | 10.13 | 182.51 |
| MCLA | 9.95 | 320.93 |
| QMI | 9.36 | 186.25 |
| EACAL | 9.74 | 205.59 |
| AdacVote | 9.22 | 172.12 |

## 6.7 Summary

We proposed a new clustering approach, OKLC, to optimize the combination of multiple kernels and Laplacians in clustering analysis. The objective of OKLC is formulated as a Rayleigh quotient function and is solved iteratively as a bi-level optimization procedure. In the simplest interface, the proposed algorithm only requires one input parameter, the cluster number $k$, from the user. Moreover, depending on user's expectation to select the most relevant sources or to evenly combine all sources, the sparseness of coefficient vector $\theta$ can be controlled via the parameter $\delta$. We proposed three variants of the OKLC algorithm and validated them on two real applications. The performance of clustering was systematically compared with a variety of algorithms and different experimental settings. The proposed OKLC algorithms performed significantly better than other methods. Moreover, the coefficients of kernels and Laplacians optimized by OKLC showed strong correlation with the rank of performance of individual data source. Though in our evaluation the $k$ values were predefined, in practical studies, the optimal cluster number can be consistently estimated from the eigenspectrum of the combined kernel Laplacian matrix.

The proposed OKLC algorithm demonstrates the advantage of combining and leveraging information from heterogeneous data structures and sources. It is potentially useful in bioinformatics and many other application areas, where there is a surge of interest to integrate similarity based information and interaction based relationships in statistical analysis and machine learning.

## References


1. Ayad, H.G., Kamel, M.S.: Cumulative voting consensus method for partitions with a variable number of clusters. IEEE Trans. PAMI 30, 160–173 (2008)
2. Bach, F.R., Lanckriet, G.R.G., Jordan, M.I.: Multiple kernel learning, conic duality, and the SMO algorithm. In: Proceedings of 21st International Conference of Machine Learning. ACM Press, New York (2004)
3. Baeza-Yates, R., Ribeiro-Neto, B.: Modern Information Retrieval. ACM Press, New York (1999)
4. Bishop, C.M.: Pattern recognition and machine learning. Springer, New York (2006)
5. Csiszar, I., Tusnady, G.: Information geometry and alternating minimization procedures. Statistics and Decisions suppl. 1, 205–237 (1984)
6. Dhillon, I.S., Guan, Y., Kulis, B.: Kernel k-means, Spectral Clustering and Normalized Cuts. In: Proceedings of the 10th ACM KDD, pp. 551–556. ACM Press, New York (2004)
7. Ding, C., He, X.: K-means Clustering via Principal Component Analysis. In: Proc. of ICML 2004, pp. 225–232. ACM Press, New York (2004)
8. Duda, R.O., Hart, P.E., Stork, D.G.: Pattern Classification, 2nd edn. John Wiley & Sons Inc., New York (2001)
9. Fred, A.L.N., Jain, A.K.: Combining multiple clusterings using evidence accumulation. IEEE Trans. PAMI 27, 835–850 (2005)
10. Girolami, M.: Mercer Kernel-Based Clustering in Feature Space. IEEE Trans. Neural Network 13, 780–784 (2002)
11. Hagen, L., Kahng, A.: New spectral methods for ratio cut partitioning and clustering. IEEE Trans. Computer-Aided Design 11, 1074–1085 (1992)
12. Hastie, T., Tibshirani, R., Friedman, J.: The Elements of Statistical Learning: Data Mining, Inference, and Prediction, 2nd edn. Springer, Heidelberg (2009)
13. Hubert, L., Arabie, P.: Comparing Partition. Journal of Classification 2, 193–218 (1985)
14. Kloft, M., Brefeld, U., Sonnenburg, S., Laskov, P., Müller, K.R., Zien, A.: Efficient and Accurate Lp-norm Multiple Kernel Learning. In: Advances in Neural Information Processing Systems 22. MIT Press, Cambridge (2009)
15. Lanckriet, G.R.G., Cristianini, N., Bartlett, P., Ghaoui, L.E., Jordan, M.I.: Learning the Kernel Matrix with Semidefinite Programming. Journal of Machine Learning Reserach 5, 27–72 (2005)
16. Liu, X., Yu, S., Janssens, F., Glänzel, W., Moreau, Y., De Moor, B.: Weighted Hybrid Clustering by Combining Text Mining and Bibliometrics on Large-Scale Journal Database. Journal of the American Society for Information Science and Technology 61, 1105–1119 (2010)

17. Mika, S., Rätsch, G., Weston, J., Schölkopf, B.: Fisher discriminant analysis with kernels. In: IEEE Neural Networks for Signal Processing IX: Proceedings of the 1999 IEEE Signal Processing Society Workshop, pp. 41–48 (1999)
18. Ng, A.Y.: On spectral clustering: Analysis and an algorithm. Advances in Neural Information Processing 14, 849–856 (2001)
19. Shawe-Taylor, J., Cristianini, N.: Kernel methods for pattern analysis. Cambridge University Press, Cambridge (2004)
20. Shi, J., Malik, J.: Normalized cuts and image segmentation. IEEE Trans. PAMI 22, 888–905 (2000)
21. Sonnenburg, S., Rätsch, G., Schäfer, C., Schölkopf, B.: Large Scale Multiple Kernel Learning. Journal of Machine Learning Research 7, 1531–1565 (2006)
22. Strehl, A., Ghosh, J.: Clustering Ensembles: a knowledge reuse framework for combining multiple partitions. Journal of Machine Learning Research 3, 583–617 (2002)
23. Suykens, J.A.K., Van Gestel, T., Brabanter, J., De Moor, B., Vandewalle, J.: Least Squares Support Vector Machines. World Scientific Press, Singapore (2002)
24. Topchy, A.: Clustering Ensembles: models of consensus and weak partitions. IEEE Trans. PAMI 27, 1866–1881 (2005)
25. Wang, F., Ding, C., Li, T.: Integrated KL(K-means-Laplacian) Clustering: A New Clustering Approach by Combining Attribute Data and Pairwise Relations. In: Proceeding of SIAM SDM 2009, pp. 38–49. SIAM Press, Philadelphia (2009)
26. Chen, J.H., Zhao, Z., Ye, J.P., Liu, H.: Nonlinear adaptive distance metric learning for clustering. In: Proceeding of ACM KDD, pp. 123–132. ACM Press, New York (2007)
27. Ye, J.P., Ji, S.H., Chen, J.H.: Multi-class discriminant kernel learning via convex programming. Journal of Machine Learning Research 40, 719–758 (2008)
28. Yu, S., Tranchevent, L.-C., Liu, X., Glänzel, W., Suykens, J.A.K., De Moor, B., Moreau, Y.: Optimized data fusion for kernel K-means clustering. Internal Report 08-200, ESAT-SISTA, K.U.Leuven, Lirias number: 242275 (2008) (submitted for publication)
29. Yu, S., Tranchevent, L.-C., De Moor, B., Moreau, Y.: Gene prioritization and clustering by multi-view text mining. BMC Bioinformatics 11, 1–48 (2010)
30. Yu, S., Falck, T., Daemen, A., Suykens, J.A.K., De Moor, B., Moreau, Y.: L2-norm multiple kernel learning and its application to biomedical data fusion. BMC Bioinformatics 11, 1–53 (2010)
31. von Luxburg, U.: A tutorial on spectral clustering. Statistics and Computing 17, 395–416 (2007)

# Chapter 7
# Weighted Multiple Kernel Canonical Correlation

## 7.1 Introduction

In the preceding chapters we have presented several supervised and unsupervised algorithms using kernel fusion to combine multi-source and multi-representation of data. In this chapter we will investigate a different unsupervised learning problem *Canonical Correlation Analysis* (CCA), and its extension in kernel fusion techniques. The goal of CCA (taking two data sets for example) is to identify the canonical variables that minimize or maximize the linear correlations between the transformed variables [8]. The conventional CCA is employed on two data sets in the observation space (original space). The extension of CCA on multiple data sets is also proposed by Kettenring and it leads to different criteria of selecting the canonical variables, which are summarized as 5 different models: sum of correlation model, sum of squared correlation model, maximum variance model, minimal variance model and generalized variance model [9]. Akaho generalizes CCA by kernel methods to find the canonical variables of data sets in the Hilbert space [1]. Bach and Jordan further propose a kernel version of CCA on multiple data sets [2], which is known as *Multiple Kernel Canonical Correlation Analysis* (MKCCA). MKCCA is also useful as a independence measure to find uncorrelated variables in the Hilbert space [2]. In this chapter, we will show that MKCCA is also useful to extract common information through maximization of the pairwise correlations among multiple data sources. A weighted extension of MKCCA can be easily derived through modifications of the objective of MKCCA [12]. The weighted MKCCA method can also be extended to out-of-sample points, which becomes important for model selection. Another important issue for MKCCA is that the problem scales up exponentially with the number of incorporated data sets and the number of samples. To make this method applicable on machines with standard CPU and memory, low rank approximation techniques based on *Incomplete Cholesky Decomposition* (ICD) and *Singular Value Decomposition* (SVD) are introduced. Moreover, for the weighted extension of MKCCA, a incremental SVD algorithm is proposed to avoid recomputing eigenvalue decomposition each time from scratch when the weights of MKCCA are updated.

S. Yu et al.: Kernel-based Data Fusion for Machine Learning, SCI 345, pp. 173–190.
springerlink.com 

## 7.2 Acronyms

The symbols and notations used in this Chapter are defined in Table 1 (in the order of appearance).

**Table 7.1** Symbols used in Chapter 7

| | | |
|---|---|---|
| $\mathbf{x}$ | $\mathbb{R}^N$ | the input data variables |
| $\mathbf{w}$ | $\mathbb{R}^N$ | the canonical variables |
| $C$ | $\mathbb{R}^N$ | the covariance matrix |
| $u,v$ | $\mathbb{N}$ | the index parameter of data sets |
| $\mathscr{O}$ | $\mathbb{R}^{N\times N}$ | the correlation among multiple sets |
| $j$ | $\mathbb{N}$ | the index of multiple sets |
| $p$ | $\mathbb{N}$ | the number of multiple sets |
| $\rho$ | $\mathbb{R}^+$ | the correlation coefficient |
| $k$ | $\mathbb{N}$ | the index of data points in each data set |
| $N$ | $\mathbb{N}$ | the number of data points |
| $\phi(\cdot)$ | $\mathbb{R}^\Phi$ | feature map |
| $\mathscr{K}(\cdot,\cdot)$ | $\mathbb{R}^\Phi \times \mathbb{R}^\Phi \to \mathbb{R}$ | kernel trick |
| $G$ | $\mathbb{R}^{N\times N}$ | centered kernel matrix |
| $\mathbf{b}$ | $\mathbb{R}^N$ | projection of data |
| $\kappa$ | $\mathbb{R}^+$ | regularization parameter |
| $\xi$ | $\mathbb{R}^+$ | weights assigned to paired correlations |
| $\theta$ | $\mathbb{R}^+$ | the decomposed weights assigned to each data sources |
| $\mathscr{W}$ | $\mathbb{R}^{pN\times pN}$ | the block diagonal matrix of weights |
| $\Omega$ | $\mathbb{R}^{pN\times pN}$ | the block matrix of paired correlations |
| $\Psi$ | $\mathbb{R}^{pN\times pN}$ | the regularized block matrix of centered kernels |
| $\psi$ | $\mathbb{R}^+$ | the normalization parameter |
| $\lambda$ | $\mathbb{R}$ | the generalized eigenvalues |
| $\mathscr{D}$ | $\mathbb{R}^{pN\times pN}$ | the Cholesky decomposition matrix of $\Psi$ |
| $\mathscr{C}$ | $\mathbb{R}^{pN\times pM}$ | the incomplete Cholesky decomposition matrix of $\Psi$, $M<N$ |
| $\alpha,\beta$ | $\mathbb{R}^N$ | eigenvectors of generalized eigenvalue problems |
| $U,\Lambda,V$ | $\mathbb{R}^{N\times N}$ | the singular value decomposition of $\mathscr{C}$, we have $\mathscr{C}=U\Lambda V^T$ |
| $E$ | $\mathbb{R}^{N\times M}$ | the orthogonal component of $U$ , $M<N$ |
| $\hat{\Lambda}$ | $\mathbb{R}^{M\times M}$ | the block component of $\Lambda$, a diagonal matrix |
| $R$ | $\mathbb{R}^{M\times M}$ | the transformed diagonal matrix |
| $\mathscr{U}$ | $\mathbb{R}^{pM\times pM}$ | the combined matrix of $U$ from multiple sets |
| $\mathscr{R}$ | $\mathbb{R}^{pM\times pM}$ | the combined matrix of $R$ from multiple sets |
| $\eta$ | $\mathbb{R}^+$ | the precision parameter of incomplete Cholesky decomposition |
| $\varepsilon$ | $\mathbb{R}^+$ | the precision parameter of singular value decomposition to select $\hat{\Lambda}$ |
| $\Delta$ | $\mathbb{R}^{pM\times pM}$ | the weights update matrix of WMKCCA |
| $\delta$ | $\mathbb{R}^+$ | the update ratio of weights $\theta$ |
| $\mathscr{A}$ | $\mathbb{R}^{pM\times pM}$ | the approximation matrix in generalized eigenvalue problem |
| $\gamma$ | $\mathbb{R}^{pM}$ | the eigenvectors of $\mathscr{A}$ |
| $\Gamma$ | $\mathbb{R}^{pM\times pM}$ | the diagonal matrix of eigenvalues of $\mathscr{A}$ |
| $\tau$ | $\mathbb{N}$ | the index parameter in incremental EVD update of WMKCCA |
| $\mathscr{T}$ | $\mathbb{R}^{pM\times pM}$ | the additive update matrix in incremental EVD update of WMKCCA |
| $t$ | $\mathbb{R}$ | the elements in matrix $\mathscr{T}$ |
| $\mathscr{B}$ | $\mathbb{R}^{pM\times pM}$ | the updated new matrix in the generalized eigenvalue problem |
| $\sigma$ | $\mathbb{R}^+$ | the kernel width of RBF kernel function |

## 7.3 Weighted Multiple Kernel Canonical Correlation

### 7.3.1 Linear CCA on Multiple Data Sets

The problem of CCA consists in finding linear relations between two sets of variables [8]. For the problem of two variables $\mathbf{x}_1$ and $\mathbf{x}_2$ with zero means, the objective is to identify vectors $\mathbf{w}_1$ and $\mathbf{w}_2$ such that the correlation between the projected variables $\mathbf{w}_1^T\mathbf{x}_1$ and $\mathbf{w}_2^T\mathbf{x}_2$ is maximized, given by

$$\max_{\mathbf{w}_1,\mathbf{w}_2} \rho = \frac{\mathbf{w}_1^T C_{\mathbf{x}_1\mathbf{x}_2}\mathbf{w}_2}{\sqrt{\mathbf{w}_1^T C_{\mathbf{x}_1\mathbf{x}_1}\mathbf{w}_1}\sqrt{\mathbf{w}_2^T C_{\mathbf{x}_2\mathbf{x}_2}\mathbf{w}_2}}, \tag{7.1}$$

where $C_{\mathbf{x}_1\mathbf{x}_1} = \mathscr{E}[\mathbf{x}_1\mathbf{x}_1^T], C_{\mathbf{x}_2\mathbf{x}_2} = \mathscr{E}[\mathbf{x}_2\mathbf{x}_2^T], C_{\mathbf{x}_1\mathbf{x}_2} = \mathscr{E}[\mathbf{x}_1\mathbf{x}_2^T]$. Extending this objective function to multiple sets of variables $\mathbf{x}_1,\ldots,\mathbf{x}_p$, one obtains the form of multiple CCA, given by

$$\max_{\mathbf{w}_1,\ldots,\mathbf{w}_p} \rho = \frac{\mathscr{O}[\mathbf{x}_1,\ldots,\mathbf{x}_p]}{\prod_{j=1}^p \sqrt{\mathbf{w}_j^T C_{\mathbf{x}_j\mathbf{x}_j}\mathbf{w}_j}}, \tag{7.2}$$

where $\mathscr{O}[\mathbf{x}_1,\ldots,\mathbf{x}_p]$ represents the correlations among multiple sets. To keep the problem analogous as the two-set one, we use the sum of correlation criterion and rewrite (7.2) as

$$\max_{\mathbf{w}_j, 1\le u<v\le m} \rho = \frac{\sum_{u,v} \mathbf{w}_u^T C_{\mathbf{x}_u\mathbf{x}_v}\mathbf{w}_v}{\prod_{j=1}^p \sqrt{\mathbf{w}_j^T C_{\mathbf{x}_j\mathbf{x}_j}\mathbf{w}_j}}. \tag{7.3}$$

As discussed in Chapter 2, the solution is found in the generalized eigenvalue problem, given by

$$\begin{bmatrix} 0 & C_{\mathbf{x}_1\mathbf{x}_2} & \cdots & C_{\mathbf{x}_1\mathbf{x}_p} \\ \vdots & \vdots & \ddots & \vdots \\ C_{\mathbf{x}_p\mathbf{x}_1} & C_{\mathbf{x}_{p-1}\mathbf{x}_1} & \cdots & 0 \end{bmatrix} \begin{bmatrix} \mathbf{w}_1 \\ \vdots \\ \mathbf{w}_p \end{bmatrix} = \rho \begin{bmatrix} C_{\mathbf{x}_1\mathbf{x}_1} & \cdots & 0 \\ \vdots & \ddots & \vdots \\ 0 & \cdots & C_{\mathbf{x}_p\mathbf{x}_p} \end{bmatrix} \begin{bmatrix} \mathbf{w}_1 \\ \vdots \\ \mathbf{w}_p \end{bmatrix}, \tag{7.4}$$

where $\rho$ is the correlation coefficient.

### 7.3.2 Multiple Kernel CCA

*Kernel Canonical Correlation Analysis* (KCCA) is a nonlinear extension of CCA using kernel methods. The data is first mapped into a high dimensional Hilbert space induced by a kernel and then the linear CCA is applied. In this way, a linear correlation discovered in the Hilbert space corresponds to a nonlinear correlation concealed in the observation space. Let us denote $\{\mathbf{x}_1^{(k)},\ldots,\mathbf{x}_m^{(k)}\}_{k=1}^N$ as $N$ data points, respectively, obtained on $p$ data sets $\mathbf{x}_1,\ldots,\mathbf{x}_p$ and $\phi_1(\cdot),\ldots,\phi_p(\cdot)$ as the feature maps from input spaces to the high dimensional Hilbert spaces. The centered kernel matrices of the $p$ data sets becomes

$$\begin{aligned} G_1 &= [\phi_1(\mathbf{x}_1^{(1)})^T - \hat{\mu}_{\phi_1};\ldots;\phi_1(\mathbf{x}_1^{(N)})^T - \hat{\mu}_{\phi_1}], \\ &\ldots \\ G_p &= [\phi_p(\mathbf{x}_p^{(1)})^T - \hat{\mu}_{\phi_p};\ldots;\phi_p(\mathbf{x}_p^{(N)})^T - \hat{\mu}_{\phi_p}]. \end{aligned} \tag{7.5}$$

The projection vectors $\mathbf{w}_1,\ldots,\mathbf{w}_p$ lie in the span of the mapped data

$$\begin{aligned} \mathbf{b}_1 &= \Phi_1 \mathbf{w}_1, \\ &\ldots \\ \mathbf{b}_m &= \Phi_p \mathbf{w}_p. \end{aligned} \tag{7.6}$$

The resulting problem of KCCA can be deduced as the analogue of the linear CCA problem on the projected data sets $\mathbf{b}_1,\ldots,\mathbf{b}_p$ in Hilbert space:

$$\max_{\mathbf{w}_j, 1\leq u<v\leq p} \rho = \frac{\sum_{u,v} \mathbf{w}_u^T C_{\Phi_u \Phi_v} \mathbf{w}_v}{\prod_{j=1}^p \sqrt{\mathbf{w}_j^T C_{\Phi_j \Phi_j} \mathbf{w}_j}}, \tag{7.7}$$

which leads to the generalized eigenvalue problem, given by

$$\begin{bmatrix} 0 & G_1G_2 & \ldots & G_1G_p \\ \vdots & \vdots & \ddots & \vdots \\ G_pG_1 & G_{p-1}G_1 & \ldots & 0 \end{bmatrix} \begin{bmatrix} \mathbf{w}_1 \\ \vdots \\ \mathbf{w}_p \end{bmatrix} = \rho \begin{bmatrix} G_1G_1 & \ldots & 0 \\ \vdots & \ddots & \vdots \\ 0 & \ldots & G_pG_p \end{bmatrix} \begin{bmatrix} \mathbf{w}_1 \\ \vdots \\ \mathbf{w}_p \end{bmatrix}, \tag{7.8}$$

where $G_j$ denotes the centered kernel matrix of the $j$-th data set which the elements $G_j(k,l)$ are obtained using the kernel trick:

$$\mathscr{K}_j(\mathbf{x}_k, \mathbf{x}_l) = \phi_j(\mathbf{x}_k)^T \phi_j(\mathbf{x}_l). \tag{7.9}$$

However, the problem in (7.8) is trivial and the non-zero solutions of generalized eigenvalue problem are $\rho = \pm 1$. To obtain meaningful estimations of canonical correlation, it needs to be regularized [2, 5, 6]. The regularization, *e.g.*, is done by the method proposed by Hardoon *et al.* [6] which results in the following regularized general eigenvalue problem:

$$\begin{bmatrix} 0 & G_1G_2 & \ldots & G_1G_p \\ \vdots & \vdots & \ddots & \vdots \\ G_pG_1 & \ldots & \ldots & 0 \end{bmatrix} \begin{bmatrix} \mathbf{w}_1 \\ \vdots \\ \mathbf{w}_p \end{bmatrix} = \rho \begin{bmatrix} (G_1 + \kappa I)^2 & \ldots & 0 \\ \vdots & \ddots & \vdots \\ 0 & \ldots & (G_p + \kappa I)^2 \end{bmatrix} \begin{bmatrix} \mathbf{w}_1 \\ \vdots \\ \mathbf{w}_p \end{bmatrix} \tag{7.10}$$

where $\kappa$ is a small positive regularization constant.

### 7.3.3 Weighted Multiple Kernel CCA

Based on the objective function in (7.7), the weighted extension of MKCCA can be formulated by employing additional weights $\xi_{u,v}$ on the pairwise correlations as

$$\max_{\mathbf{w}_j, 1 \le u < v \le p} \rho = \frac{\sum_{u,v} \xi_{u,v} \mathbf{w}_u^T C_{\Phi_u \Phi_v} \mathbf{w}_v}{\prod_{j=1}^p \sqrt{\mathbf{w}_j^T C_{\Phi_j \Phi_j} \mathbf{w}_j}} \tag{7.11}$$

where $\xi_{u,v}$ is the scalar weight of the correlation between $\mathbf{x}_u$ and $\mathbf{x}_v$. Let us denote the generalized eigenvalue problem in (7.10) as the form of $\Omega\alpha = \lambda\Psi\alpha$, the weights of KCCA is decomposed as an additional positive definite matrix $\mathscr{W}$ multiplying at the left and right side of the matrix $\Omega$, given by

$$\mathscr{W}\Omega\mathscr{W}\alpha = \lambda\Psi\alpha \tag{7.12}$$

where

$$\mathscr{W} = \begin{bmatrix} \theta_1 I & 0 & \dots & 0 \\ 0 & \theta_2 I & \dots & 0 \\ \vdots & \vdots & \dots & \vdots \\ 0 & 0 & \dots & \theta_p I \end{bmatrix},$$
$$\Omega = \begin{bmatrix} 0 & G_1 G_2 & \dots & G_1 G_p \\ G_2 G_1 & 0 & \dots & G_2 G_p \\ \vdots & \vdots & \ddots & \vdots \\ G_p G_1 & G_{p-1} G_1 & \dots & 0 \end{bmatrix},$$
$$\Psi = \begin{bmatrix} (G_1 + \kappa I)^2 & 0 & \dots & 0 \\ 0 & (G_2 + \kappa I)^2 & \dots & 0 \\ \vdots & \vdots & \ddots & \vdots \\ 0 & 0 & \dots & (G_p + \kappa I)^2 \end{bmatrix},$$
$$\sum_{j=1}^p \theta_j = p, \quad \xi_{u,v} = \psi \theta_u \theta_v, \quad \psi = \frac{1}{\sum_{1 \le u < v \le p} \theta_u \theta_v}.$$

In the formulation above, the weights of pairwise correlation $\xi$ in the objective function (7.11) are decomposed as the weights $\theta$ assigned on each data set. The sum of $\theta$ is constrained to keep the mean value as 1. $\psi$ is a normalization parameter to make the sum of $\xi$ equal to 1. This normalization constant $\psi$ only affects the solution of eigenvalue but does not affect the eigenvector solution.

## 7.4 Computational Issue

### 7.4.1 *Standard Eigenvalue Problem for WMKCCA*

To solve the computational burden, we follow the transformation presented in [2] to rewrite the generalized eigenvalue problem in (7.12) as

$$[\mathscr{W}\Omega\mathscr{W} + \Psi]\alpha = (\lambda + 1)\Psi\alpha. \tag{7.13}$$

The problem of finding the maximal generalized eigenvalues in (7.12) is equivalent to finding the minimal generalized eigenvalues in (7.13) because if the generalized eigenvalues in (7.12) are given by

$$\lambda_1, -\lambda_1, \ldots, \lambda_a, \lambda_a, 0, \ldots, 0,$$

then the corresponding generalized eigenvalues in (7.13) are

$$1 + \lambda_1, 1 - \lambda_1, \ldots, 1 + \lambda_a, 1 - \lambda_a, 1, \ldots, 1.$$

Since $\Psi$ is already regularized thus it can be decomposed as $\Psi = \mathscr{D}^T\mathscr{D}$, defining $\beta = \mathscr{D}\alpha$, and $\mathscr{K}_\kappa = \mathscr{W}\Omega\mathscr{W} + \Psi$, the problem in (7.13) can be transformed as follows:

$$\begin{aligned} \mathscr{K}_\lambda \alpha &= \lambda^\sharp \Psi \alpha \\ \mathscr{K}_\lambda \alpha &= \lambda^\sharp \mathscr{C}^T \mathscr{C} \alpha \\ \mathscr{D}^{-T} \mathscr{K}_\kappa \mathscr{D}^{-1} \beta &= \lambda^\sharp \beta \end{aligned} \tag{7.14}$$

Notice that $\Psi$ is a positive definite matrix in a block diagonal form, thus we have

$$\mathscr{D}^T = \mathscr{D} = \Psi^{1/2} = \begin{bmatrix} G_1 + \kappa I & \ldots & 0 \\ \vdots & \ddots & \vdots \\ 0 & \ldots & G_p + \kappa I \end{bmatrix}. \tag{7.15}$$

Replace (7.15) in (7.14), the problem is written in the form of a standard eigenvalue problem, given by

$$\begin{bmatrix} I & \ldots & \theta_1\theta_p I r_\kappa(G_1) r_\kappa(G_p) \\ \vdots & \ddots & \vdots \\ \theta_1\theta_p I r_\kappa(G_p) r_\kappa(G_1) & \ldots & I \end{bmatrix} \beta = \lambda^\sharp \beta, \tag{7.16}$$

where

$$r_\kappa(G_j) = G_j(G_j + \kappa I)^{-1}, \;\; j = 1, \ldots, p.$$

If eigenvalues $\lambda^\sharp$ and eigenvectors $\beta$ are solved from (7.16), the eigenvalues and eigenvectors of problem (7.12) is $\lambda^\sharp$ and $\mathscr{C}^{-1}\beta$. More formally, the eigenvectors $\alpha_j$ of problem (7.10) are equivalent to

$$\alpha_j = (G_j + \kappa I)^{-1}\beta_j, \quad j = 1, ..., p. \tag{7.17}$$

## 7.4.2 Incomplete Cholesky Decomposition

As introduced in [2], the full rank $N \times N$ centered kernel matrix $G_j$ can be factorized as $G_j \approx \mathscr{C}\mathscr{C}^T$ by ICD, where $\mathscr{C}_j$ is in low rank $N \times M_j$ ($M_j \le N$) matrix. Applying SVD on $\mathscr{C}_j$ one obtains $N \times M_j$ matrix $U_i$ with orthogonal columns and $M_j \times M_j$ diagonal matrix $\Lambda_j$, given by

$$G_j \approx \mathscr{C}_j\mathscr{C}_j^T = U_j\Lambda_jV_j^T(U_j\Lambda_jV_j^T)^T = U_j\Lambda_j^2U_j^T. \tag{7.18}$$

Denote $E_j$ as the orthogonal complement of $U_j$, $(U_jE_j)$ is then a full rank $N \times N$ matrix, $G_j$ is given by

$$G_j \approx U_j\Lambda_j^2U_j^T = [U_jE_j]\begin{bmatrix}\hat{\Lambda}_j & 0\\ 0 & 0\end{bmatrix}[U_jE_j]^T. \tag{7.19}$$

For regularized matrices in (7.10), one obtains:

$$r_\kappa(G_j) \approx [U_jE_j]\begin{bmatrix}R_j & 0\\ 0 & 0\end{bmatrix}[U_jE_j]^T = U_jR_jU_j^T, \tag{7.20}$$

where $R_j$ is the diagonal matrix obtained from the diagonal matrix $\hat{\Lambda}_j$ by transformation $r_j^{(k)} = (\hat{\lambda}_j^{(k)})/(\hat{\lambda}_j^{(k)} + \kappa)$ to its elements, where $r_j^{(k)}$ are elements in $R_j$ and $\hat{\lambda}_j^{(k)}$ are the diagonal elements in $\hat{\Lambda}_j$. Replacing (7.16) with (7.20), decomposing (7.16) as

$$\mathscr{U}\mathscr{R}\mathscr{U}^T\beta = \lambda^\sharp\beta, \tag{7.21}$$

where

$$\mathscr{U} = \begin{bmatrix}U_1 & \dots & 0\\ \vdots & \ddots & \vdots\\ 0 & \dots & U_p\end{bmatrix},$$
$$\mathscr{R} = \begin{bmatrix}I & \dots & \zeta_1\zeta_pIR_1U_1^TU_pR_p\\ \vdots & \ddots & \vdots\\ \zeta_1\zeta_pIR_pU_p^TU_1R_1 & \dots & I\end{bmatrix}. \tag{7.22}$$

Since $\mathscr{R}$ is derived from a similar matrix transformation, the eigenvalues in (7.21) are equivalent to (7.16), moreover, the eigenvectors of the low rank approximation is related to the full rank problem by the following transformation:

$$
\begin{aligned}
\mathscr{U}\mathscr{R}\mathscr{U}^T\beta &= \lambda^\sharp\beta \\
\mathscr{R}\mathscr{U}^T\beta &= \lambda^\sharp\mathscr{U}^T\beta \\
\mathscr{R}\gamma &= \lambda^\sharp\gamma
\end{aligned} \tag{7.23}
$$

Once we obtained the eigenvector $\gamma_j$ in low rank approximation problem (7.23), we can recover the full rank solution in (7.16) by $\beta_j = U_i\gamma_j$. Furthermore, the generalized eigenvector $\alpha_i$ of the original problem is computed as (7.17), so we have:

$$\alpha_j \approx (G_j + \kappa I)^{-1} U_j \gamma_j \tag{7.24}$$

We have several parameters involved in WMKCCA computation: $\kappa$ as the regularization parameter, $\eta$ as the precision parameter for ICD, $\varepsilon$ as the cut value of eigenvalues determining the size of $U_i$ and $\lambda_i$ in SVD of $G_i$. Among these parameters, $\eta$ and $\varepsilon$ are easy to select by numerical experiments because they monotonically control the precision of ICD and SVD. The regularization parameter $\kappa$ is often determined empirically.

### 7.4.3 *Incremental Eigenvalue Solution for WMKCCA*

Based on the weighted problem in (7.12), the update of weights in WMKCCA is equivalent to multiplying an update matrix $\Delta$ at the left and right sides of the WMKCCA formulation, given by

$$\Delta \mathscr{W} \Omega \mathscr{W} \Delta \alpha = \lambda \Psi \alpha \tag{7.25}$$

where

$$\Delta = \begin{bmatrix} \delta_1 & 0 & \dots & 0 \\ 0 & \ddots & \ddots & \vdots \\ \vdots & & \ddots & 0 \\ 0 & \dots & 0 & \delta_p \end{bmatrix}, \tag{7.26}$$

and $\delta_1, \dots, \delta_p$ are the update ratios of weights corresponding to $\theta_1, \dots, \theta_p$. Following the analog steps from (7.13) to (7.16), the standard *eigenvalue decomposition* (EVD) problem with updated weights is in the form:

$$\mathscr{A}\gamma = \lambda^\sharp\gamma, \tag{7.27}$$

where

$$\mathscr{A} = \begin{bmatrix} I & \dots & \delta_1\delta_p\theta_1\theta_p I R_1 U_1^T U_p R_p \\ \vdots & \ddots & \vdots \\ \delta_1\delta_p\theta_1\theta_p I R_p U_p^T U_1 R_1 & \dots & I \end{bmatrix}. \tag{7.28}$$

For simplicity let us denote the matrix in (7.28) before weight updating as $\mathscr{A}_{old}$, and the one after updating as $\mathscr{A}_{new}$. We denote $E = \mathscr{A}_{new} - \mathscr{A}_{old}$. For the weights updated problem, we are interested in solving $\mathscr{A}_{new}\gamma = \lambda^{\sharp}\gamma$ by reusing the solution obtained on $\mathscr{A}_{old}$. Thus, during weight update we could avoid redoing the EVD each time from scratch. Suppose we already have solution as

$$\gamma_\tau \Gamma_\tau \gamma_\tau^T = \mathscr{A}_\tau, \tag{7.29}$$

where $\tau$ is the index parameter for the iteration. Then the updated problem is equivalent to adding $\mathscr{T}$ on both sides of equation, given by

$$\begin{aligned} \gamma_\tau \Gamma_\tau \gamma_\tau^T + \mathscr{T} &= \mathscr{A}_\tau + \mathscr{T} \\ \gamma_\tau (\Gamma_\tau + \gamma_\tau^T \mathscr{T} \gamma_\tau) \gamma_\tau^T &= \mathscr{A}_{\tau+1} \\ \gamma_\tau \mathscr{B} \gamma_\tau^T &= \mathscr{A}_{\tau+1} \end{aligned} \tag{7.30}$$

In (7.28) the weight updating only affects the off-diagonal elements of the matrix. Moreover, the constraints of weights matrix in (7.12) bounds the update weights $\delta_1, ..., \delta_p$ within a certain scope. Especially, for small scale updates, these values are close to 1. Thus the matrix $\mathscr{T}$ is by

$$\mathscr{T} = \begin{bmatrix} 0 & t_{1,2} & \cdots & t_{1,p} \\ \vdots & & \ddots & \vdots \\ t_{p,1} & t_{p,2} & \cdots & 0 \end{bmatrix}, \tag{7.31}$$

where

$$t_{u,v} = (\delta_u \delta_v - 1)\theta_u \theta_v I R_u U_u^T U_v R_v. \tag{7.32}$$

As shown, $\mathscr{T}$ only has non-zero values at off-diagonal positions and most of the elements are close to 0. Because $\gamma_\tau$ is a unitary matrix, and $\gamma_\tau^T \mathscr{T} \gamma_\tau$ is also a sparse matrix with most of the off diagonal elements are close to 0, therefore, the matrix $\mathscr{B}$ in (7.30) is a nearly diagonal matrix which can be solved efficiently by iterative EVD algorithms. Through this transformation, we can have a "warm start" solution of weights by storing the previous EVD solution and computing the EVD solution of the new problem incrementally.

## 7.5 Learning from Heterogeneous Data Sources by WMKCCA

WMKCCA extracts common information among multiple heterogeneous data sets. Given a group of objects represented in different sources, the inter-relationships among these objects may contain some intrinsic patterns. Through WMKCCA, the correlations are calculated in the Hilbert space thus complex intrinsic patterns among multiple sources may be revealed. When multiple sources are combined, WMKCCA has the flexibility to leverage the importance of correlations among several data sources. The projections obtained in canonical spaces are also useful for machine learning applications. To illustrate this, we propose an learning framework based on WMKCCA. As illustrated in Figure 7.1, the proposed framework integrates WMKCCA with supervised learning where the validation data and test data are projected to the embedding of the training data by out-of-sample projection [3]. In this thesis, the model for WMKCCA is selected by cross-validating the machine learning performance on the validation set so that the parameters of kernel function and the weights assigned on correlations are optimized.

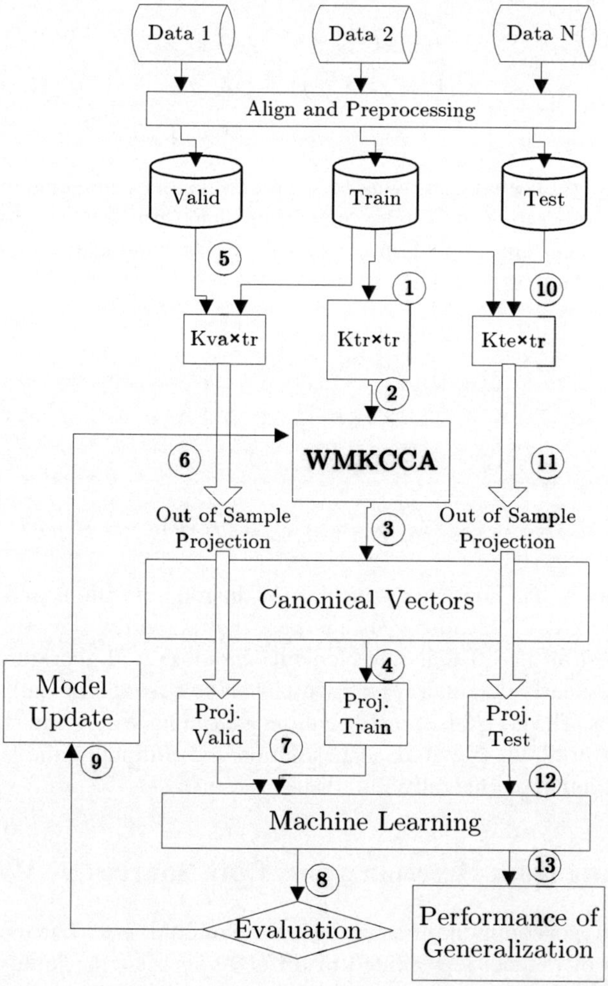


**Fig. 7.1** A framework of learning by WMKCCA on heterogeneous data sources. The numbers in the circle represent the sequence of the workflow.

## 7.6 Experiment

### 7.6.1 *Classification in the Canonical Spaces*

We adopted the two UCI pattern recognition data sets, pen-based recognition of handwritten digits and optical recognition of handwritten digits, as mentioned in Chapter 3. As abbreviation, we denote them as PenData and OptData respectively. Both data sets have 10 digit labels from 0 to 9. PenData has 16 input attributes measured from 0 to 100 and OptData has 64 attributes measured from 0 to 16. We extracted 3750 samples (375 samples for each digit) from the training part of both data sets, 80% of them were used for training and 20% were used for validation. We used their original test data as test set (3498 in PenData, 1797 in OptData). We applied RBF functions on both data sets and the kernel widths are selected as the mean of the sample covariance (for PenData $\sigma = 97$, for OptData $\sigma = 13$). We created a third data set by transforming the label information of data as a kernel matrix. Firstly, the vector of class labels was coded as $N \times 10$ matrix $L$ where the $i$-th column represents the label of the $i-1$-th digit. Then, the label matrix was transformed into a kernel matrix by linear function $LL^T$. Therefore, in our training step we used three $3000 \times 3000$ kernel matrices, denoted as $K_{pen}, K_{opt}, K_{label}$ respectively.

We applied a centroid classification approach on the projected data in the canonical spaces which treats each set of digits as a cluster and calculates the centroid as the mean prototype in canonical space. When a new sample is presented for classification, the Euclidean distances from the new point to all the cluster centroids are calculated and the label is assigned to the new data as the cluster with the shortest distance. To obtain the validation performance and test performance, the validation data and test data are firstly projected to the canonical spaces of the training data with out-of-sample projection [3], then classified by the centroid method. The accuracy of the classification is evaluated by calculating the percentage of the correctly classified data in all digit labels.

We benchmarked the performance in the validation sets with different weights, using the incremental EVD method presented in Section 7.4.3. The weights on $K_{pen}, K_{opt}, K_{label}$ are denoted as $\theta_{pen}, \theta_{opt}, \theta_{label}$ and they are optimized in a grid search from 0.1 to 2.9 with the step 0.1. We also benchmarked the dimensionality of the canonical space by evaluating the performance of classification. For each validation, the data is projected to the subspace spanned by 10, 100 and 500 canonical vectors respectively. The weights and the size of the canonical space are selected by the average classification accuracies on the two validation data sets. Then the test data is projected to the WMKCCA model parameterized by the optimal weights and the optimal canonical space. The performance of the test set is evaluated by comparing the classification labels and the true labels. Furthermore, we compared the performance obtained by WMKCCA learning with the results reported by other methods in the literature. The results obtained by KCCA, MKCCA are also compared. As shown in Table 7.2, the performance obtained by the simple classification method applied in the canonical space is comparable to the best results reported in the literature.

**Table 7.2** Classification accuracy obtained on the test data. By 10-fold cross-validation, the weights in WMKCCA are set as $\theta_{pen} = 1.3, \theta_{opt} = 1.3, \theta_{label} = 0.4$ and the dimensionality of the canonical space is set to 500. As shown, the results of WMKCCA, MKCCA and KCCA are comparable to the performance obtained by some complicated classification methods. However, the performance of WMKCCA, MKCCA and KCCA is not significantly different from each other. The performance of other algorithms is mean values reported in the literature. According to the comparison of the mean values, the proposed CCA approaches might perform better than the linear SVM. As mentioned, the kernel matrices in CCA approaches are all constructed by linear function, so the improvements should ascribe to the effects of canonical projections obtained from data fusion.

| METHODS | PenData | OptData | Notes |
|---|---|---|---|
| WMKCCA | **0.9794** | **0.9716** | See text |
| MKCCA | **0.9766** | **0.9688** | Equal weights |
| KCCA | **0.9783** | **0.9711** | 3750 training |
| Linear SVM | 0.9494 | 0.9602 | |
| RBF DDA SVM [11] | 0.9708 | 0.9722 | |
| SVM Ensemble [10] | N/A | 0.9783 | 45 SVMs |
| MLP [11] | 0.9703 | 0.9517 | |
| kNN [11] | 0.9771 | 0.9649 | k=3 |
| Bayes+PCA [7] | 0.9763 | 0.9694 | |

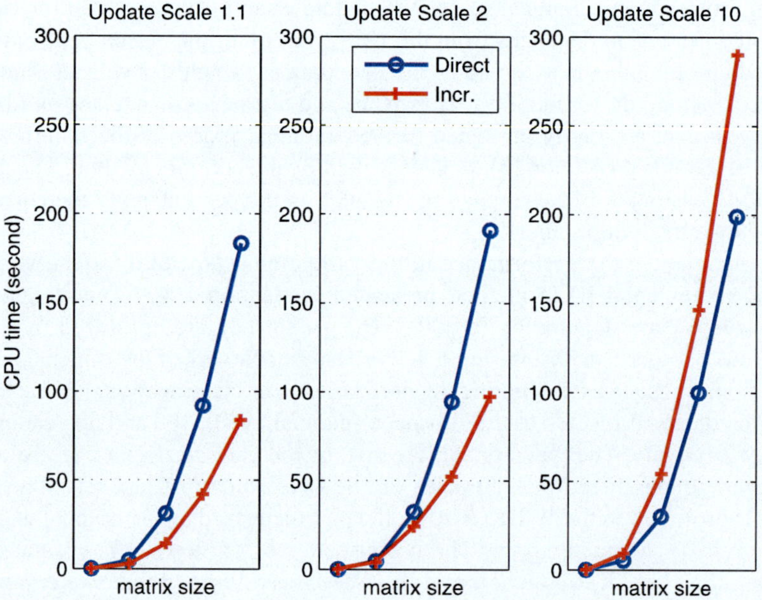


**Fig. 7.2** Comparison of CPU time used by direct EVD method and incremental EVD method in WMKCCA computation

### 7.6.2 Efficiency of the Incremental EVD Solution

We also compared the direct EVD approach and the incremental EVD algorithm in the WMKCCA experiments with different matrix sizes and update scales. The benchmark did not take into account the ICD and SVD processes but only compared the CPU time of solving the eigenvalue problem in (7.23) between the direct method and the incremental method. For the incremental method, the CPU time of calculating $E$, $T$, the EVD of $T$ and the multiplication process of calculating the canonical vectors were all taken into account. We modified the scale of the problem by increasing the size of training set as 100, 500, 1000, 1500 and 2000 data samples. The linear kernels were constructed on two data sets and fed to the WMKCCA algorithm parameterized as ($\eta = 0.9$, $\kappa = 0.1$, $\varepsilon = 0.9$). After ICD and SVD, the matrix $\mathscr{A}$ in (7.27) had the size of 160, 701, 1323, 1911 and 2483 respectively. We also adjusted the scale of the weight update parameter $v$, which is denoted as the ratio between the new weight and the old weight in $\Delta$ as (7.26). As discussed, when $\delta$ is close to 1, the off-diagonal elements in matrix $E$ become sparse. When $\delta$ is much larger than 1, $E$ is not necessarily a sparse off-diagonal matrix, thus the assumption of incremental EVD does not hold. To demonstrate this, we compared three $\delta$ values: 1.1, 2, and 10 which respectively represents a weak update, a moderate update and a strong update of weights in WMKCCA model. For the problem of 3 data sets, when the update scale is set to 2 and 10, the mean value of $\theta$ does not necessarily equal to 1 hence the constraint in (7.12) does not hold. The experiment is conducted on a desktop PC with Intel Core 2 1.86GHz CPU and 2G memory. The software package for simulation is MATLAB 2006a. The EVD algorithm is implemented by the *eig* function in MATLAB, which is based on QR method. As shown in Figure 7.2, the incremental EVD algorithm significantly reduces the CPU time when the weight update is small and moderate. However, when the update is too large, the matrix $\mathscr{B}$ is not close to diagonal thus the effect of incremental algorithm is overwhelmed by the additional cost paid on matrix multiplication.

### 7.6.3 Visualization of Data in the Canonical Spaces

Similar to the kernel CCA visualization method presented in [4] on single data sets, we could also visualize the PenData and OptData in the canonical spaces learned by WMKCCA. We present a series of figures visualizing all 300 training points in the space spanned by the 1st and 2nd canonical variate. By adjusting the weights assigned to different data sources, we are able to change the patterns of data projected in the canonical spaces. Notice that in Figure 7.3 and Figure 7.4, when the weights assigned to all sources are set to 1, WMKCCA reduces to MKCCA and the projections of PenData and OptData become very similar. The similarity can be judged by the relative positions of different digits. It is also can be noticed that the grouping of digits in PenData is more tight than the OptData. The also explains why in the previously mentioned classification results that the performance of model built on PenData is better than the one built on Optdata. On PenData (Figure 7.3), if we increase the weight assigned to PenData (the $v_1$ value) and decrease the weight assigned to OptData (the $v_2$ value), the WMKCCA projection becomes very similar to the KCCA projections of PenData illustrated in Figure 7.5, which is obtained by applying KCCA on the PenData alone. Notice that the relative positions among digit groups and their shapes are almost exactly the same. Analogously, as shown in Figure 7.4, if we increase the weighted assigned to OptData (the $v_2$ value) and decrease the PenData weight ($v_1$), the WMKCCA projection of OptData becomes similar to the KCCA projection of OptData presented in Figure 7.4.

By analyzing the visualization of projections in canonical spaces, we find that the weights in WMKCCA leverage the effect data sources during the construction of the canonical spaces. In other words, WMKCCA provides a flexible model to determine the canonical relationships among multiple data sources and this flexibility could be utilized in machine learning and data visualization.

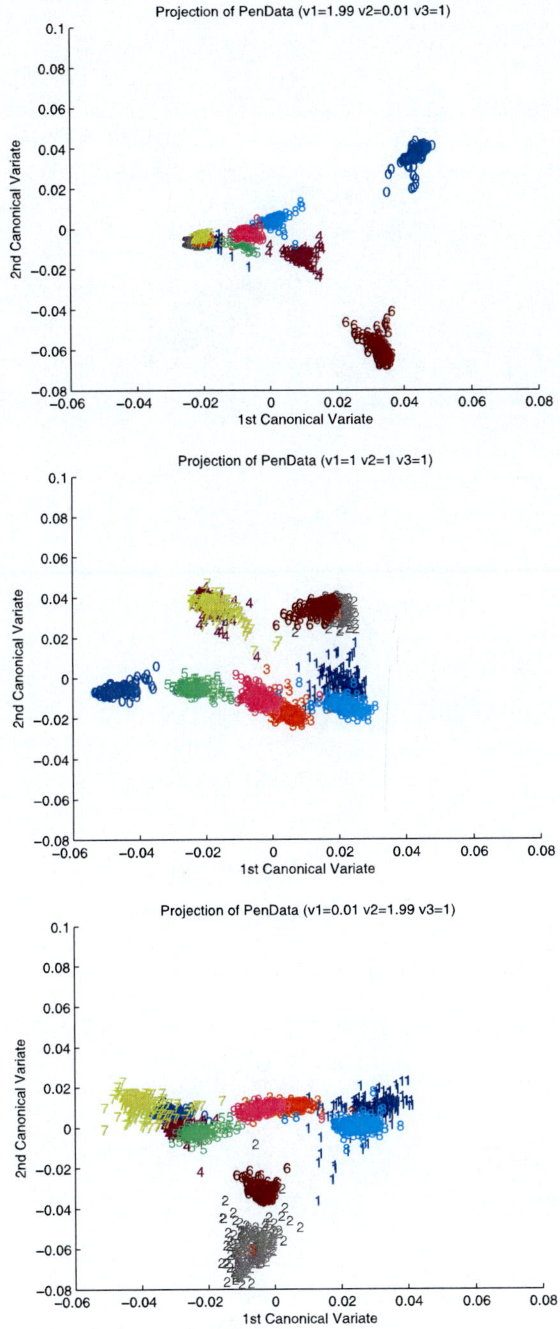


**Fig. 7.3** Visualization of Pendata in the canonical spaces obtained by WMKCCA. The v1, v2, v3 values in the figures are respectively the weights assigned to pen data, optical data, and the label data.

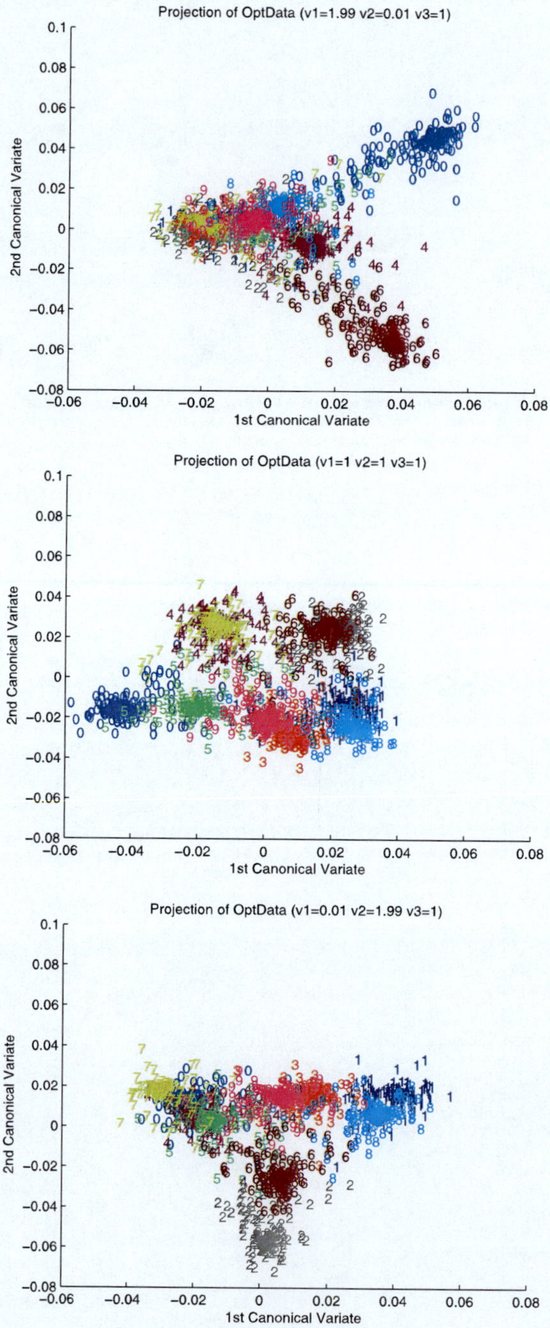


**Fig. 7.4** Visualization of Optdata in the canonical spaces obtained by WMKCCA. The v1, v2, v3 values in the figures are respectively the weights assigned to pen data, optical data, and the label data.

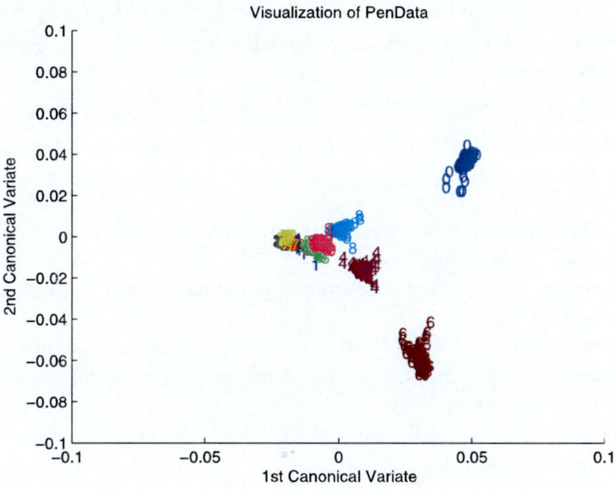


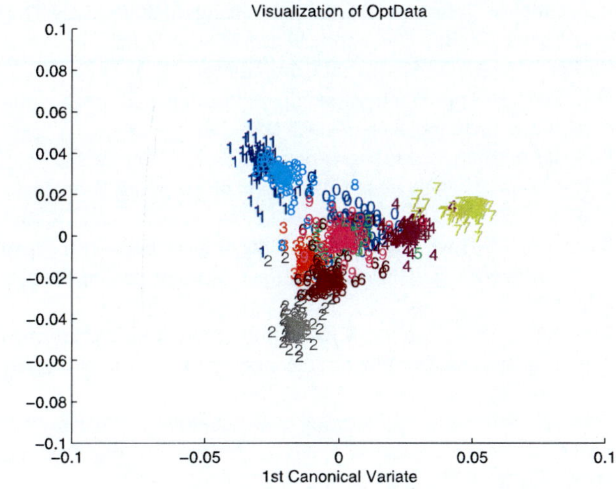


**Fig. 7.5** Visualization of PenData (top) and Optdata (bottom) independently in the canonical spaces obtained by KCCA

## 7.7 Summary

In this chapter we proposed a new weighted formulation of kernel CCA on multiple sets. Using low rank approximation and incremental EVD algorithm, the WMKCCA is applicable in machine learning problems as a flexible model to extract common information among multiple data sources. We tried some preliminary experiments to demonstrate the effect. The CCA based data fusion is in a different paradigm than the MKL method introduced in previous chapters. The multiple data sources are not merged, whereas the merit of data fusion relies in the subspace spanned by the canonical vectors. In machine learning, the projections of data in these canonical spaces may be useful and more significant to detect the common underlying patterns reside in multiple sources.

## References


1. Akaho, S.: A kernel method for canonical correlation analysis. In: Proc. of the International Meeting of Psychometric Society 2001 (2001)
2. Bach, F.R., Jordan, M.I.: Kernel independent component analysis. Journal of Machine Research 3, 1–48 (2003)
3. Bengio, Y., Paiement, J.F., Vincent, P.: Out-of-Sample Extensions for LLE, Isomap, MDS, Eigenmaps, and Spectral Clustering. Advances in Neural Information Processing Systems 15, 177–184 (2003)
4. Chang, Y.I., Lee, Y.J., Pao, H.K., Lee, M., Huang, S.Y.: Data visualization via kernel machines. Technical report, Institute of Statistical Science, Academia Sinica (2004)
5. Gretton, A., Herbrich, R., Smola, A., Bousquet, O., Scholkopf, B.: Kernel methods for measuring independence. Journal of Machine Research 6, 2075–2129 (2005)
6. Hardoon, D.R., Shawe-Taylor, J.: Canonical Correlation Analysis: An Overview with Application to Learning Methods. Neural Computation 16, 2639–2664 (2004)
7. Holmstrom, L., Hoti, F.: Application of semiparametric density estimation to classification. In: Proc. of ICPR 2004, pp. 371–374 (2004)
8. Hotelling, H.: Relations between two sets of variates. Biometrika 28, 321–377 (1936)
9. Kettenring, J.R.: Canonical analysis of several sets of variables. Biometrika 58, 433–451 (1971)
10. Kim, H.C., Pang, S., Je, H.M., Bang, S.Y.: Pattern classification using support vector machine ensemble. In: Proc. of 16th International Conference on Pattern Recognition, pp. 160–163 (2002)
11. Oliveira, A.L.I., Neto, F.B.L., Meira, S.R.L.: Improving rbfdda performance on optical character recognition through parameter selection. In: Proc. of 18th International Conference on Pattern Recognition, pp. 625–628 (2004)
12. Yu, S., De Moor, B., Moreau, Y.: Learning with heterogeneous data sets by Weighted Multiple Kernel Canonical Correlation Analysis. In: Proc. of the Machine Learning for Signal Processing XVII. IEEE, Los Alamitos (2007)

# Chapter 8
# Cross-Species Candidate Gene Prioritization with MerKator

## 8.1 Introduction

In modern biology, the use of high-throughput technologies allows researchers and practicians to quickly and efficiently screen the genome in order to identify the genetic factors of a given disorder. However these techniques are often generating large lists of candidate genes among which only one or a few are really associated to the biological process of interest. Since the individual validation of all these candidate genes is often too costly and time consuming, only the most promising genes are experimentally assayed. In the past, the selection of the most promising genes relied on the expertise of the researcher, and its *a priori* opinion about the candidate genes. However, *in silico* methods have been developed to deal with the massive amount of complex data generated in the post-sequence era. In the last decade, several methods have been developed to tackle the gene prioritization problem (recently reviewed in [13]). An early solution was proposed by Turner *et al.* who proposed POCUS in 2003 [14]. POCUS relies on Gene Ontology annotations, InterPro domains, and expression profiles to identify the genes potentially related to the biological function of interest. The predictions are made by matching the Gene Ontology annotations, InterPro domains and expression profile of the candidate genes to the ones of the genes known to be involved in the biological function of interest. The system favors the candidate genes that exhibit similarities with the already known genes. Most of the proposed prioritization methods also rely on this 'guilt-by-association' concept. Several methods rely solely on text-mining but nowadays most of the novel methods combine textual information with experimental data to leverage the effect between reliability and novelty.

Most of the existing approaches are restricted to integrating information in a single species. Recently, people have started to collect phylogenetic evidences among multiple species to facilitate the prioritization of candidate genes. Chen *et al.* proposed 'ToppGene' that performs prioritization for human based on human data (*e.g.*, functional annotations, proteins domains) as well as mouse data (*i.e.*, phenotype

S. Yu et al.: Kernel-based Data Fusion for Machine Learning, SCI 345, pp. 191–205.
springerlink.com 

data) [3]. Through an extensive validation, they showed the utility of mouse phenotype data in human disease gene prioritization. Hutz *et al.* [5] have developed CANDID, an algorithm that combines cross-species conservation measures and other genomic data sources to rank candidate genes that are relevant to complex human diseases. In their approach, they adopted the NCBI's HomoloGene database that analyzes genes from 18 organisms and collects homologs using amino acid and DNA sequences. Liu *et al.* have investigated the effect of adjusting gene prioritization results by cross-species comparison. They identified the ortholog pairs between Drosophila melanogaster and Drosophila pseudoobscura by BLASTP and used this cross-species information to adjust the rankings of the annotated candidate genes in D. melanogaster. They report that a candidate gene with a lower score in the main species (D. melanogaster) may be re-ranked higher if it exhibits strong similarity to orthologs in coding sequence, splice site location, or signal peptide occurrence [6]. According to the evaluation on the test set of 7777 loci of D. melanogaster, the cross-species model outperforms other single species models in sensitivity and specificity measures. Another related method is String developed by von Mering *et al.* [15]. String is a database that integrates multiple data sources from multiple species into a global network representation.

In this paper, we present MerKator, whose main feature is the cross-species prioritization through genomic data fusion over multiple data sources and multiple species. This software is developed on the Endeavour data sources [1, 12] and a kernel fusion novelty detection methodology [4]. To our knowledge, MerKator is one of the first real bioinformatics softwares powered by kernel methods. It is also one of the first cross-species prioritization softwares freely accessible online. In this chapter, we present and discuss the computational challenges inherent to such implementation. We also present a benchmark analysis, through leave-one-out cross-validation, that shows the efficiency of the cross-species approach.

## 8.2 Data Sources

The goal of MerKator is to facilitate the understandings of human genetic disorders using genomic information across organisms. MerKator identifies the homologs of *Homo sapiens*, denoted as the main organism, in four reference organisms: *Mus musculus*, *Rattus norvegicus*, *Drosophila melanogaster*, and *Caenorhabditis elegans*. The identification is based on NCBI's HomoloGene [8, 16, 17], which provides the mapping of homologs among the genes of 18 completely sequenced eukaryotic genomes. For each gene in each organism, MerKator stores the homolog pair with the lowest ratio of amino acid differences (the *Stats-prot-change* field in HomoloGene database). MerKator incorporates 14 genomic data sources in multiple species for gene prioritization. The complete list of the data sources adopted in the current version is presented in Table 8.1.

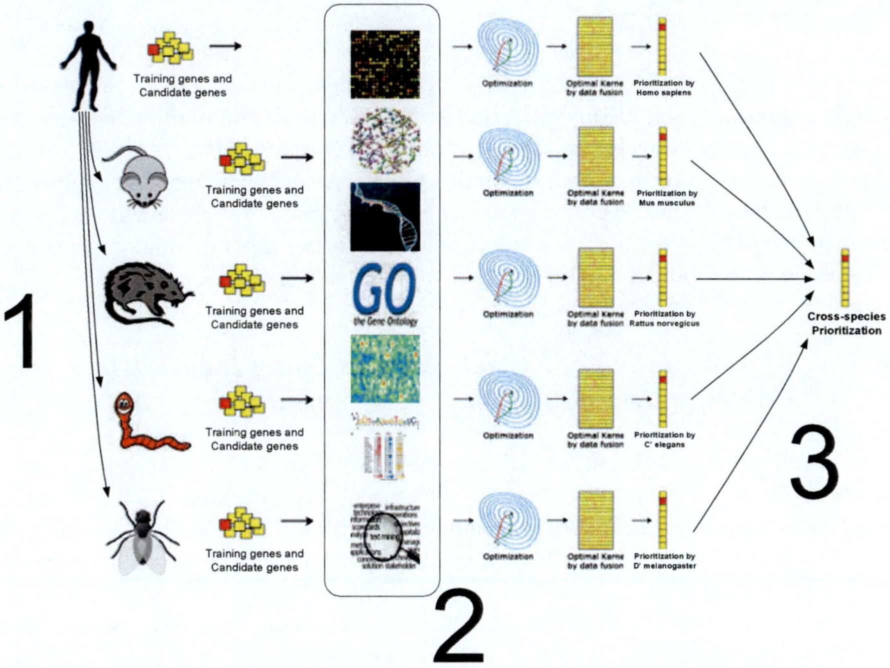


**Fig. 8.1** Conceptual overview of Endeavour MerKator software. The clip arts of species obtained from Clker.com by Brain Waves LLC.

**Table 8.1** Genomic data sources adopted in Endeavour MerKator

| data source | H.sapiens | M.musculus | R.norvegicus | D.melanogaster | C.elegans |
|---|---|---|---|---|---|
| Annotation GO | √ | √ | √ | √ | √ |
| Annotation-Interpro | √ | √ | √ | √ | √ |
| Annotation-EST | √ | √ | | | |
| Sequence-BLAST | √ | √ | √ | √ | √ |
| Annotation-KEGG | √ | √ | √ | √ | √ |
| Expression-Microarray | √ | √ | √ | √ | √ |
| Annotation-Swissprot | | √ | √ | √ | √ |
| Text | √ | | | | |
| Annotation-Phenotype | | | | √ | |
| Annotation-Insitu | | | | √ | |
| Motif | √ | | | | |
| Interaction-Bind | √ | | | | |
| Interaction-Biogrid | √ | | | √ | √ |
| Interaction-Mint | | | | √ | √ |

## 8.3 Kernel Workflow

MerKator applies 1-SVM method [4, 9, 11] to obtain prioritization scores within a single organism. Then the prioritization scores obtained from multiple species are integrated using a Noisy-Or model. As mentioned, MerKator is a real bioinformatics software powered by kernel methods therefore many challenges are tackled in its design and implementation. Considering the efficiency of kernel methods implemented in real full-genomic scale application, MerKator separates the program into the offline process and the online process to improve its efficiency.

### 8.3.1 *Approximation of Kernel Matrices Using Incomplete Cholesky Decomposition*

The main computational burden is the kernel computation of various data sources in the full genomic scale, especially for the data that is represented in high dimensional space, such as Gene Ontology annotations, gene sequences, and text-mining among others. To tackle this difficulty, MerKator manages all the kernel matrices in an offline process using a Matlab-Java data exchange tool. In Matlab, the tool retrieves the genomic data from the databases and construct the kernel matrices. The kernel matrices of the full genomic data may be very large so it is not practical to handle them directly in the software. To solve this, we decompose all the kernel matrices with ICD (Incomplete Cholesky Decomposition), thus the dimensions of the decomposed kernel matrices are often smaller than the original data. In MerKator, the precision of the ICD is set as 95% of the matrix norm, given by

$$\frac{||K-K'||_2}{||K||_2} \leq 0.05, \tag{8.1}$$

where $K$ is the original kernel matrix, $K'$ is the approximated kernel matrix as the inner product of the ICD matrix. In this way the computational burden of kernel calculation is significantly reduced as the computation of the inner product of the decomposed matrices. The Matlab-Java tool creates Java objects on the basis of decomposed kernel matrices in Matlab and stores them as serialized Java objects. The kernel computation, its decomposition and the Java object transformation are computationally intensive processes, and so they are all executed offline. For the online process, MerKator loads the decomposed kernel matrices from the serialized java objects, reconstructs the kernel matrices and solve the 1-SVM MKL optimization problem to prioritize the genes as already described in De Bie *et al.* [4]. Then the prioritization results are displayed on the web interface. In contrast with the offline process, the online process is less computational demanding and the complexity is mainly determined by the $d$ number of training genes ($O(d^3)$). In our implementation, the optimization solver is based on the Java API of MOSEK[2] that shows satisfying performance.

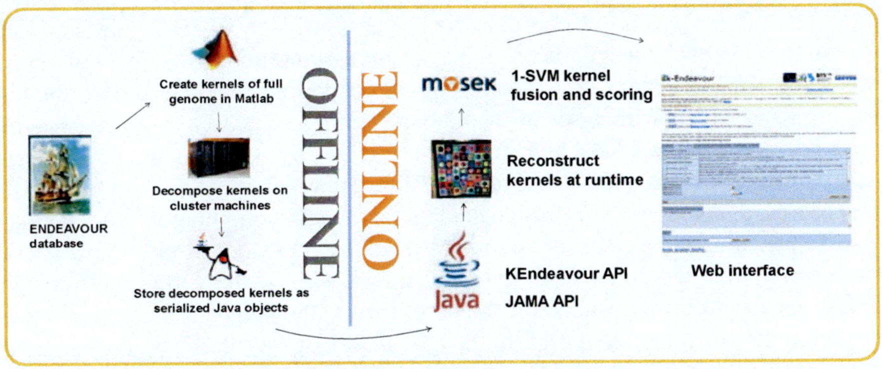


**Fig. 8.2** Separation of offline and online processes in MerKator

### *8.3.2 Kernel Centering*

In MerKator, when the prioritization task involves data set of the full genomic size, some trivial operations become quite inefficient. To control the number of false positive genes in 1-SVM, De Bie *et al.* suggest a strategy to center the kernel matrices that contain both the training genes and the test genes on the basis of the *iid* assumption. As mentioned in the work of Shawe-Taylor and Cristianini [10], the kernel centering operation expressed on the kernel matrix can be written as

$$\hat{K} = K - \frac{1}{l}\mathbf{1}\mathbf{1}^T K - \frac{1}{l}K\mathbf{1}\mathbf{1}^T + \frac{1}{l^2}(\mathbf{1}^T K\mathbf{1})\mathbf{1}\mathbf{1}^T, \tag{8.2}$$

where $l$ is the dimension of $K$, $\mathbf{1}$ is the all 1s vector, $T$ is the vector transpose. Unfortunately, when the task is to prioritize the full genomic data, centering the full genome kernel matrices becomes very inefficient. For MerKator, we use a strategy based on the split of the full genomic data into smaller subsets. Let us assume that the full genome data contains $\mathscr{N}$ genes, and is split into several subsets containing $\mathscr{M}$ genes. Instead of centering the kernel matrix sizes of $\mathscr{N} \times \mathscr{N}$, we center the kernel matrix of size $\mathscr{A} \times \mathscr{A}$, where $\mathscr{A}$ is the number of genes in the union of the $\mathscr{M}$ candidate genes with the training genes. Because $\mathscr{M}$ is smaller than $\mathscr{N}$, for each centered kernel matrix MerKator obtains the prioritization score of $\mathscr{M}$ candidate genes, so it need to iterate multiple times (denoted as $k$, which is the smallest integer larger than $\frac{\mathscr{N}}{\mathscr{M}}$) to calculate the scores of all the $\mathscr{N}$ candidate genes. According to the *iid* assumption, if $\mathscr{M}$ is large enough then centering the kernel matrix of size $\mathscr{A} \times \mathscr{A}$ is statistically equivalent to centering the kernel matrix of the full genome, thus the prioritization scores obtained from the $k$ iterations can precisely approximate the values obtained when centering the full genome data. Therefore, we may compare the prioritization scores of the $\mathscr{N}$ genes even if they are obtained from different centered matrices, and thus we can prioritize the full genome. All these assumptions come to one non-trivial question: how to select the appropriate $\mathcal{M}$ to trade off between the reliability of *iid* assumption and the computational efficiency? In MerKator, $\mathcal{M}$ is determined via experiments conducted on the text mining data source with a set of 20 human training genes. First, a prioritization model is built and the 22743 human genes are scored by centering the linear kernel matrix of the full genome. The obtained values are regarded as the true prioritization scores, denoted as $f$. We also calculate the overall computation time, denoted as $t$. To benchmark the effect of $\mathcal{M}$, we try 10 different values from 1000 to 10000. In each iteration, the 20 training genes are mixed with $\mathcal{M}$ randomly selected candidate genes and the prioritization scores of the candidate genes are computed by centering the small kernel matrix. In the next iteration, we select $\mathcal{M}$ new candidate genes until all the 22743 genes are prioritized. The prioritization scores obtained by centering this small kernel matrix are denoted as $f'$, and the computation time is also compared. The difference (error) between the prioritization scores obtained in these two approaches represents how well the $\mathcal{M}$ candidate genes approximates the *iid* assumption of the full genome, and is given by

$$e = \frac{||f - f'||_2}{||f||_2}. \tag{8.3}$$

We use this difference to find the optimal $\mathcal{M}$. According to the benchmark result presented in Table 8.2, large $\mathcal{M}$ values lead to small error but take much longer time for the program to center the kernel matrix. In MerKator, we set the $\mathcal{M}$ to 4000, which represents a balance between a low error ($e < 0.05$) and a fast computing time (16 times faster than centering the full genome).

**Table 8.2** The approximation error and the computational time of using $\mathcal{M}$ randomly selected genes in kernel matrix centering

| $\mathcal{M}$ | $e$ | time(seconds) |
|---|---|---|
| 22,743 | 0 | 11969 |
| 10000 | 0.015323701440092 | 4819.2 |
| 9000 | 0.022324135658694 | 3226.3 |
| 8000 | 0.028125449554702 | 2742.0 |
| 7000 | 0.048005271001603 | 2135.2 |
| 6000 | 0.041416998355952 | 1638.2 |
| 5000 | 0.048196878290559 | 1117.0 |
| **4000** | **0.045700854755551** | **745.40** |
| 3000 | 0.087474107488752 | 432.76 |
| 2000 | 0.098294618397952 | 191.95 |
| 1000 | 0.136241837096454 | 72.34 |

### 8.3.3 *Missing Values*

In bioinformatics applications, clinical and genomic datasets are often incomplete and contain missing values. This is also true for the genomic data sources that underly MerKator, for which a significant number of genes are missing. In MerKator, the missing gene profiles are represented as zeros in the kernel matrices mainly for computational convenience. However, zeros still contain strong information so that they may lead to imprecise prioritization scores. In MerKator, kernel matrices are linearly combined to create the global kernel that is used to derive the prioritization scores. In order to avoid relying on missing data for this calculation (and therefore to favor the well studied genes), we use a strategy illustrated in Supplementary Figure 8.2 to combine kernel matrices with missing values. This strategy is similar to what is done within Endeavour. For a given candidate gene, only the non-missing based scores are combined to calculate the overall score. The combined kernel matrix obtained by this strategy is still a valid positive semi-definite kernel and thus the obtained prioritization scores only rely on the non-missing information.

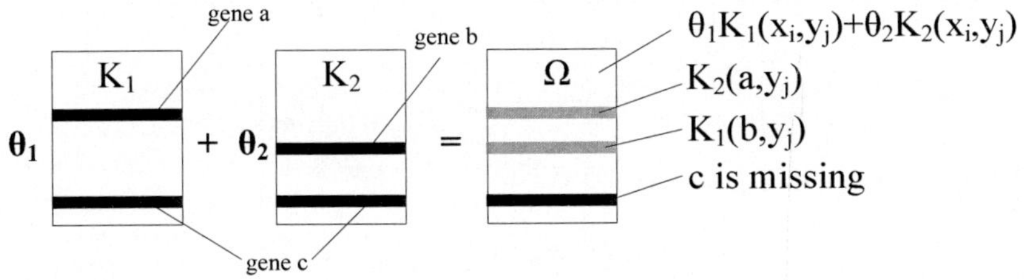


**Fig. 8.3** Kernel fusion with missing values. Suppose gene *a* and gene *c* are missing in the first kernel $K1$; gene *b* and gene *c* are missing in the second kernel $K2$; $\theta_1$, $\theta_2$ are the kernel coefficients. In the combined kernel $\Omega$, we fill the missing values of gene *a* by the non-missing information in $K2$ and gene *b* by the non-missing information in $K1$. The gene *c* is missing in all data sources so it will not be prioritized in MerKator. For the case with more than 2 data sources, the coefficients of missing values are recalculated by considering non-missing sources only.

## 8.4 Cross-Species Integration of Prioritization Scores

MerKator first uses the 1-SVM algorithm to prioritize genes in a single species, and then adopts a Noisy-Or model [7] to integrate prioritization scores from multiple species. We assume the scenario of cross-species prioritization as depicted in Figure 8.1. Similar to Endeavour, MerKator takes a machine learning approach by building a disease-specific model on a set of disease relevant genes, denoted as *training set*, then that model is used to rank the candidate genes, denoted as *candidate set*, according to their similarities to the model.

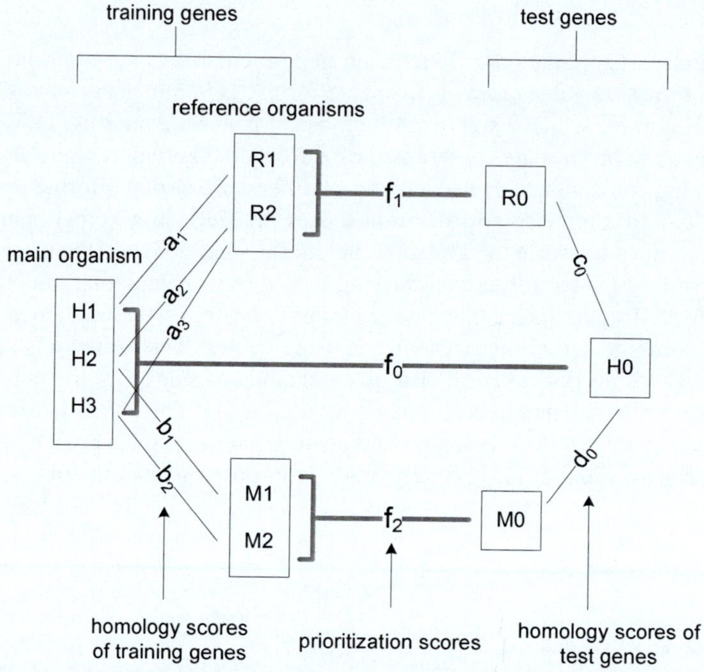


**Fig. 8.4** Integration of cross-species prioritization scores

Some fundamental notions in cross-species gene prioritization are illustrated in Figure 8.4. Suppose in the main organism we specify $N_1$ number of human genes as $\{H_1, ..., H_{N_1}\}$ and MerKator obtains the corresponding training sets in reference species rat and mouse. The training set of rat contains $N_2$ genes as $\{R_1, ..., R_{N_2}\}$ and the training set of mouse has $N_3$ genes as $\{M_1, ..., M_{N_3}\}$. Note that MerKator always selects the homolog with the highest similarity ratio of sequence, so it is a *many-to-one* mapping thus $N_2$, $N_3$ are always smaller or equal to $N_1$. We define the homolog scores between the training sets of human and rat as $a_1, ..., a_{N2}$; Similarly, the homolog scores between human and mouse training sets are $b_1, ..., b_{N3}$. For the candidate set, each candidate gene of human is mapped to at most one rat gene and one mouse gene, where the homolog score is respectively denoted as $c_0$ and $d_0$. The homolog genes and the associated scores are all obtained from the NCBI HomoloGene database (release 63). To calculate the cross-species prioritization score, we introduce a set of utility parameters as follows.

We denote $h_1$ and $h_2$ as the parameters describing the quality of the homology, given by:

$$h_1 = \min\{c_0, \text{median}(a_1, a_2, ..., a_{N2})\},$$
$$h_2 = \min\{d_0, \text{median}(b_1, b_2, ..., b_{N3})\}.$$

$z_1$ and $z_2$ are denoted as the parameters describing the ratio of the number of homologs in the reference organism with the number of genes in the main organism, given by

$$z_1 = \frac{N_2}{N_1}, \quad z_2 = \frac{N_3}{N_1}. \tag{8.4}$$

Next, we denote $f_0$ as the prioritization score of candidate gene $H_0$ ranked by the training set $\{H_1, ..., H_{N1}\}$; denote $f_1$ as the score of reference candidate gene $R_0$ prioritized by the reference training set $\{R_1, ..., R_{N2}\}$ and $f_2$ as the score of $M_0$ ranked by the set $\{M_1, ..., M_{N3}\}$. The raw prioritization scores obtained by 1-SVM are in the range of $[-1, +1]$, thus we scale them into $[0, +1]$. The adjustment coefficient $adj$ is defined as:

$$adj = 1 - \prod_{\text{organism } i} (1 - h_i z_i f_i). \tag{8.5}$$

The $adj$ coefficient combines information from multiple species by the Noisy-Or model. A larger $adj$ means there is strong evidence from the homologs that the candidate gene is relevant to the model. Considering the case one may want to eliminate the homolog bias, we further correct the $adj$ parameter, denoted as $adj^+$, given by

$$adj^+ = \begin{cases} \text{median}(\{adj\}), & \text{if } j \text{ has no homolog} \\ 1 - \{\prod_{\text{organism } i} (1 - h_i z_i f_i)\}^{\frac{1}{k}} & \text{if } j \text{ has homolog(s)}. \end{cases} \tag{8.6}$$

The first case of equation (8.6) means that when gene $j$ has no homolog related, its $adj^+$ score equals to the median value of the set $\{adj\}$ that contains the adjustment values of the genes that have at least one homolog gene. In the second case, when there are $k$ number of homologs mapped to gene $j$, we use the $k$-th exponential root removes the additional bias of the prioritization score caused by the multiple homologs.

This coefficient $adj^+$ is used to adjust $f_0$, the prioritization score of the main organism, and we have tried two different versions as follows:

$$\text{human non-special: } f_{\text{cross-species}} = 1 - (1 - f_0)(1 - adj^+), \tag{8.7}$$

and

$$\text{human special: } f_{\text{cross-species}} = 1 - \frac{(1 - f_0)(2 - adj^+)}{2}. \tag{8.8}$$

The *human non-special* version considers the human prioritization score as equivalent to the homology evidence and combines them again using the Noisy-Or function. In contrast, the *human special* version only adjusts the human prioritization score with the homology evidence by average. In the Noisy-Or integration, the cross-species score is boosted up if either the main species or the homology evidence shows good prioritization score. In the average score integration, the cross-species score compromises between the homology evidence and the main species score, and is only boosted up if both of them are good.

## 8.5 Software Structure and Interface

This section presents the interface of the software and explains how MerKator works. Using MerKator, a prioritization can be prepared in 4 steps (see Figure 8.5). In the first step, the user has to define the main organism, it will be the reference organism and will be used to input the training and candidate genes. In addition, the user can select other organisms to use, the corresponding species specific data sources will then be included further in the analysis. If no other organism is selected, the results are only based on the main organism data sources. In a second step, the training genes are inputed. Genes from the main organism can be input using various gene identifiers (*e.g.*, EnsEMBL, gene name, EntrezGene) or even pathway identifiers from KEGG or Gene Ontology. In addition, for human, an OMIM entry number can be inputted. Genes are loaded into the system using the 'Add' button. In the third step, the data sources to be used are selected by checking the corresponding boxes. By default, only the data sources of the main organism are displayed and the program is automatically selecting the corresponding data sources in the reference organisms when available. To have a full control on the data sources, the user must enter the advanced mode by clicking the dedicated button. Using the advanced mode, data sources from other organisms can be selected individually. In the fourth step, the candidate genes to prioritize are inputted. The user has two possibilities,

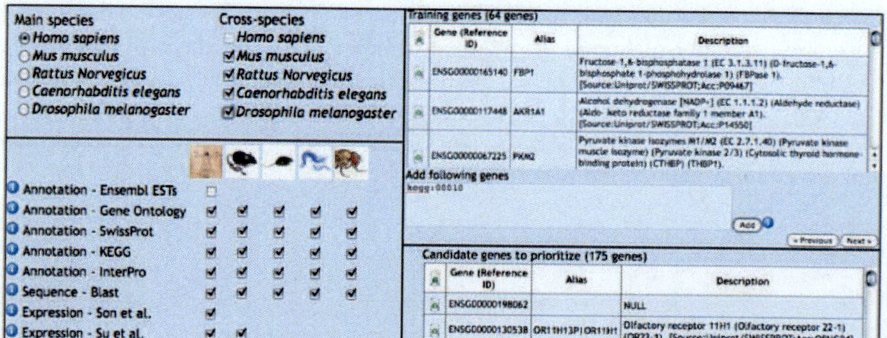


**Fig. 8.5** Typical MerKator workflow. This includes the species selection step (first step - top left), the input of the training genes (second step - top right), the selection of the data sources (third step - bottom left) and the selection of the candidate genes (fourth step - bottom right). Screenshots were taken from our online web server.

either use the whole genome (in case the results are returned by e-mail) or input a subset of the genome (in case results are displayed in the web interface). For the latter, the method is similar to the second step, but genomic regions can also be inputted (*e.g.*, band q11 on chromosome 22, or region 100k - 900k on chromosome 19). The prioritization can be launched from this panel.

## 8.6 Results and Discussion

The chapter introduces MerKator, a novel gene prioritization software based on kernel methods that can perform cross-species prioritization in five organisms (human, rat, mouse, fruit fly and worm). Compared to the previous approaches, our method differs by the number of organisms combined (current prioritization approaches focus either on mouse or fruit fly as well as by the information that is combined (current prioritization approaches focus on conservation or expression data). The String approach of von Mering *et al.* is, to some respect, similar to our approach. The main differences with our method are, first, that String predicts novel interactions but does not perform prioritization, and, second, that String relies mostly on its text-mining component while we aim at integrating several genomic data sources (including but not restricted to text-mining).

To improve the efficiency of MerKator, we tackle the kernel computational challenges of full genomic data from multiple aspects. First, most of the computation was done offline and performed only once, restricting the case specific online computation to a strict minimum. Second, the prioritization of the full genome utilizes some approximation techniques such as incomplete Cholesky decomposition, kernel centering in the subsets of genome, and missing value processing to improve its feasibility and efficiency. Based on these efforts, MerKator is able to integrate all the adopted data sources from five species and prioritize the full human genome within 15 minutes.

We have developed a Noisy-Or based method to integrate the scores from multiple species into a global score. This Noisy-Or based method integrates the species specific scores by taking into account the strength of the homology between the corresponding species. The use of a Noisy-Or based method is motivated by the fact that an excellent prioritization score obtained in one species should be enough to obtain an overall excellent score, which other measures such as the average would not allow. We have developed two scoring schemes termed *human special* and *human non-special*. The first one assumes that the source species (human in our case) is the main organism and that the other species are only used to adapt the score obtained in the main organism. The contribution of the other species is a half in total (the other half is the main organism score). The second solution is however relying on the hypothesis that all the species can contribute evenly, the main organism is not distinguished from the others. We have implemented and analyzed the two methods. In addition, we have implemented and tested two formula for the adjustment coefficient, $adj$ and $adj^+$, to account for the differences in number of homolog genes. We have observed that the *adj* coefficient can introduce a bias towards the genes that have multiple homologs as compared to the genes that do not have any homologs. Either the homolog genes are still unknown or there is no homolog in any of the other species and therefore the gene is a human specific gene. In both cases, there is no rationale behind the bias and the gene should get the same chance to rank high than the other genes.

As a proof of concept, we have benchmarked MerKator with five biological pathways using a leave-one-out cross-validation procedure. The four pathways were derived from Gene Ontology and contain a total of 37 genes. The validation was performed using all data sources except Gene Ontology and all five species. In this case, we have used the both formulas *human special* and *human non special*. The Area Under the ROC Curve (AUC) is used as an indicator of the performance. We obtained a global AUC of 89.38% for the cross-species model (human non-special), while the model based on human data alone obtains a smaller AUC of 82.64% (see Figure 8.6). For four out of the five pathways, the cross-species model performs better than the human specific model although significance is not reached given the low number of genes per pathways (between 6 and 9). For the remaining pathway (GO:0008199), the two models achieve similar performance. This is because the human only performance is already too high to allow any significant improvement (AUC >99.9%). These results indicate that our cross-species model is conceptually valid and that reference organism genomic data can enhance the performance of human gene prioritization.

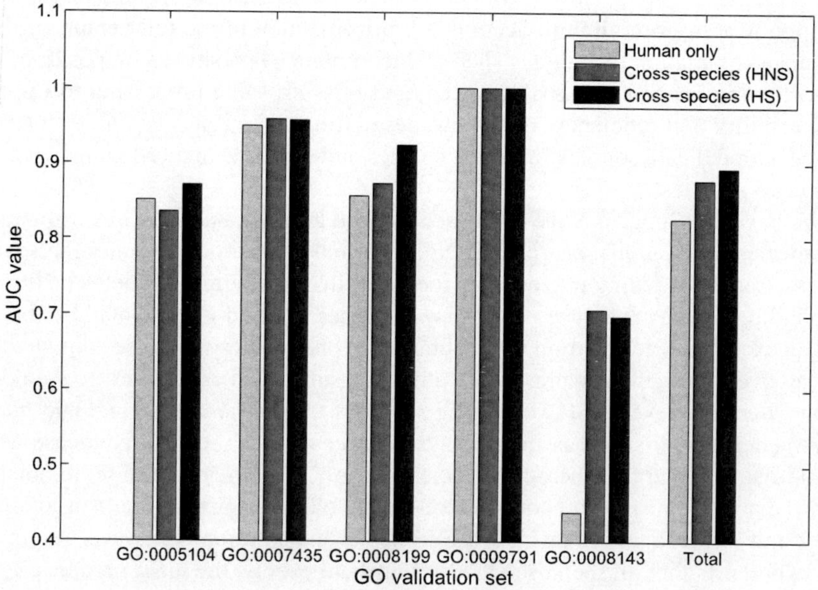


**Fig. 8.6** Benchmark results on five GO pathways using human only data sources (grey bars) and our cross-species data integration model (black bars)

## 8.7 Summary

We present MerKator, a software that combines cross-species information and multiple genomic data sources to prioritize candidate genes. The software is developed using the same databases adopted in Endeavour, but is equipped with a kernel fusion technique and a cross-species integration model. To embed kernel methods in a real and large scale bioinformatics application, we have tackled several computational challenges that are mentioned and discussed in this paper. Our approach may be concluded with the following three aspects:

- *Combining evidences from multiple species.* We proposed a Noisy-Or model to combine prioritization scores from multiple organisms. The issue of multiple species prioritization is complicated, which may involve many factors such as the size of training set, the selection of data sources, the number of relevant homologies, and so on. Considering so many factors, it is difficult to make statistical hypothesis, or estimate the data model for the final prioritization score. Thus our approach alternatively avoids the assumption about the data model of prioritization scores and calculates it using support vector machines. The integration methods are adjusted in the blackbox and the outputs are validated with benchmark data until satisfying performance is obtained.
- *User friendly interface.* Gene prioritization softwares are oriented to a specific group of computational biologists and medical researchers, therefore we designed an user friendly interface that is similar to Endeavour's web interface, and that does not require advanced mathematical skills to be used (configuration of the 1-SVM and the integration models are transparent to the end users). The results of full genomic cross-species prioritization are either directly returned or stored on the server and delivered to the end-user by e-mail messages depending on the number of candidate genes. When receiving the email notice, the users can either upload the prioritization results and display them in MerKator or download the results in XML format to extract the relevant information by themselves.
- *Near optimal solution.* The performance of kernel-based algorithms is strongly affected by the selection of hyper-parameters, such as the parameter of kernel function, or the regularization parameter. The optimal parameters should be selected by cross-validation, which may not be always feasible for a software oriented for biologists and medical researchers. Kernel fusion techniques allow developers to preselect the kernel parameters empirically. The overall performance does not rely on a single kernel parameter, so even when the optimal parameter is not involved, the fusion procedure still can leverage among several near optimal parameters and provides a near optimal result. For real applications, the 1% difference of performance is not so critical to the end users. In most cases, a successful application prefers much the speed of solution than the very optimality of the parameter or the model.

Future work includes, but is not restricted to, the inclusion of more species and more data sources, the development of new modules to enhance even further the performance of our kernel based prioritization algorithm, the parallelization of computational methods to incorporate more data sources and more species, and the application to real biological problems, for instance through the integration of MerKator into research workflows.

## References


1. Aerts, S., Lambrechts, D., Maity, S., Van Loo, P., Coessens, B., De Smet, F., Tranchevent, L.C., De Moor, B., Marynen, P., Hassan, B., Carmeliet, P., Moreau, Y.: Gene prioritization through genomic data fusion. Nature Biotechnology 24, 537–544 (2006)
2. Andersen, E.D., Andersen, K.D.: The MOSEK interior point optimizer for linear programming: an implementation of the homogeneous algorithm. In: High Perf. Optimization, pp. 197–232. Kluwer Academic Publishers, New York (2000)
3. Chen, J., Xu, H., Aronow, B.J., Jegga, A.G.: Improved human disease candidate gene prioritization using mouse phenotype. BMC Bioinformatics 8, 392 (2007)
4. De Bie, T., Tranchevent, L.C., Van Oeffelen, L., Moreau, Y.: Kernel-based data fusion for gene prioritization. Bioinformatics 23, i125–i123 (2007)
5. Hutz, J.E., Kraja, A.T., McLeod, H.L., Province, M.A.: CANDID: a flexible method for prioritizing candidate genes for complex human traits. Genetic Epidemiology 32, 779–790 (2008)
6. Liu, Q., Crammer, K., Pereira, F.C.N., Roos, D.S.: Reranking candidate gene models with cross-species comparison for improved gene prediction. BMC Bioinformatics 9, 433 (2008)
7. Pearl, J.: Probabilistic Reasoning in Intelligent Systems: Networks of Plausible Inference. Morgan Kaufmann, San Francisco (1988)
8. Sayers, E.W., Barrett, T., Benson, D.A., Bolton, E., Bryant, S.H., Canese, K., Chetvernin, V., Church, D.M., Dicuccio, M., Federhen, S., Feolo, M., Geer, L.Y., Helmberg, W., Kapustin, Y., Landsman, D., Lipman, D.J., Lu, Z., Madden, T.L., Madej, T., Maglott, D.R., Marchler-Bauer, A., Miller, V., Mizrachi, I., Ostell, J., Panchenko, A., Pruitt, K.D., Schuler, G.D., Sequeira, E., Sherry, S.T., Shumway, M., Sirotkin, K., Slotta, D., Souvorov, A., Starchenko, G., Tatusova, T.A., Wagner, L., Wang, Y., Wilbur, W.J., Yaschenko, E., Ye, J.: Database resources of the National Center for Biotechnology Information. Nucleic Acids Research 38, D5–D16 (2010)
9. Schölkopf, B., Platt, J.C., Shawe-Taylor, J., Smola, A.J., Williamson, R.C.: Estimating the support of a high-dimensional distribution. Neural Computation 13, 1443–1471 (2001)
10. Shawe-Taylor, J., Cristianini, N.: Kernel methods for pattern analysis. Cambridge University Press, Cambridge (2004)
11. Tax, D.M.J., Duin, R.P.W.: Support vector domain description. Pattern Recognition Letter 20, 1191–1199 (1999)
12. Tranchevent, L., Barriot, R., Yu, S., Van Vooren, S., Van Loo, P., Coessens, B., De Moor, B., Aerts, S., Moreau, Y.: ENDEAVOUR update: a web resource for gene prioritization in multiple species. Nucleic Acids Research 36, W377–W384 (2008)
13. Tranchevent, L.C., Capdevila, F.B., Nitsch, D., De Moor, B., Causmaecker, P., Moreau, Y.: A guide to web tools to prioritize candidate genes. Briefings in Bioinformatics 11, 1–11 (2010)

14. Turner, F.S., Clutterbuck, D.R., Semple, C.A.M.: POCUS: mining genomic sequence annotation to predict disease genes. Genome Biology 4, R75 (2003)
15. von Mering, C., Jensen, L.J., Snel, B., Hooper, S.D., Krupp, M., Foglierini, M., Jouffre, N., Huynen, M.A., Bork, P.: STRING: known and predicted protein-protein associations, integrated and transferred across organisms. Nucleic Acids Research. 33, D433–D437 (2005)
16. Wheeler, D.L., Barrett, T., Benson, D.A., Bryant, S.H., Canese, K., Chetvernin, V., Church, D.M., DiCuccio, M., Edgar, R., Federhen, S., Geer, L.Y., Helmberg, W., Kapustin, Y., Kenton, D.L., Khovayko, O., Lipman, D.J., Madden, T.L., Maglott, D.R., Ostell, J., Pruitt, K.D., Schuler, G.D., Schriml, L.M., Sequeira, E., Sherry, S.T., Sirotkin, K., Souvorov, A., Starchenko, G., Suzek, T.O., Tatusov, R., Tatusova, T.A., Wagner, L., Yaschenko, E.: Database resources of the National Center for Biotechnology Information. Nucleic Acids Research 34, D173–D180 (2006)
17. Wheeler, D.L., Church, D.M., Lash, A.E., Leipe, D.D., Madden, T.L., Pontius, J.U., Schuler, G.D., Schriml, L.M., Tatusova, T.A., Wagner, L., Rapp, B.A.: Database resources of the National Center for Biotechnology Information. Nucleic Acids Research 29, 11–16 (2001)

# Chapter 9
# Conclusion

The exquisite nature of combining various senses in human cognition motivated our approach to incorporate multiple sources in data mining. The research described in this book covers a number of topics which are relevant to supervised and unsupervised learning by kernel-based data fusion. The discussion of these topics were distinguished in four different aspects: theory, algorithm, application and software.

In the theoretical part, we reviewed the mathematical objectives of several popular supervised and unsupervised techniques in machine learning. The Rayleigh quotient objective was shown as equivalent as the objective of the kernel fusion method based on LS-SVM MKL. The equivalence of the objectives enabled us to plug the LS-SVM MKL in unsupervised learning problems to develop various new algorithms. To tackle the computational burden of large scale problems, we investigated several optimization techniques to simplify the problem. One of the main findings was that the SIP formulation of LS-SVM MKL is very efficient, being much faster than, while showing identical solution as other formulations on the benchmark data. We also investigated the optimization of different norms in the dual problem of Multiple Kernel Learning (kernel fusion). The selection of norms in kernel fusion yields different characteristics of optimal weights assigned to multiple sources. In particular, the $L_2$-norm yields non-sparse weights, which were shown empirically better than the sparse weights obtained from $L_\infty$-norm in some real biomedical applications.

In the algorithmic part, we presented three unsupervised kernel fusion algorithms. The first two were proposed for clustering analysis: the OKKC method combines multiple kernels for clustering, which is suitable for attribute-based data integration; the OKLC method combines kernels and Laplacian in clustering, which was proposed for fusing attribute-based data and interaction-based data. The third algorithm, WMKCCA, extends the conventional unweighted canonical correlation method to a weighted version.

As one of the main characteristics of the book, we applied all the proposed algorithms in real applications. We applied the 1-SVM MKL algorithm to combine multiple text mining views and multiple genomic data sources in disease candidate gene prioritization. We also proposed a framework, including the OKKC, the OKLC, and

S. Yu et al.: Kernel-based Data Fusion for Machine Learning, SCI 345, pp. 207–208.
springerlink.com 

a set of other popular algorithms, to combine heterogeneous data sources in clustering analysis. This clustering framework was applied to partition disease relevant genes using multi-view text mining data. It was also applied in a scientometrics application to combine text mining and bibliometric data to obtain mapping fields of science and technology.

Finally, we solidified our efforts in kernel-based gene prioritization as the Endeavour MerKator software. MerKator is a real kernel-based software designed to prioritize candidate genes in terms of combining evidence from multiple species and multiple genomic data sources. It handles large amount of data at the full genome scale. Many challenges were encountered in the design and development stage of MerKator and some of our strategies are presented. The software will be freely accessible online for biologists and bioinformaticians.

The theories, algorithms, applications and softwares presented in this book provide an interesting perspective for kernel-based data fusion especially in Bioinformatics and text mining. Moreover, the obtained results are promising to be applied and extended to many other relevant fields as well.

# Index